Assessing Public Health Needs in a Lower Middle Income Country

This book demonstrates a methodology for assessing public health needs in communities experiencing environmental sanitation inadequacies. Centring on a case study of the Republic of Cameroon, the findings represent the starting point of a campaign to implement a comprehensive sanitation infrastructure.

Based on an assessment report undertaken by ARCHIVE Global, an international non-profit organization focusing on the link between health and housing, this book:

- Explores and establishes a causal relationship between the built environment and its impacts on public health
- Uses the United Nations' Sustainable Development Goals as a benchmark for highlighting issues and challenges with sanitation infrastructure projects
- Provides lessons for communities around the world facing environmental health issues similar to those Cameroon's Idenau Municipality deals with.

This book is intended for environmental health professionals, academics, and policy makers, be they domestic to the African region or multinational practitioners. Donor countries, the likes of the United States and European nations, will also value the book's advocacy for interventions in the built environment and current public health impacts.

Sarah Ruel-Bergeron, RA, Director of Projects and Development, ARCHIVE Global

Sarah Ruel-Bergeron is a licensed architect and the Executive Director at ARCHIVE Global, a non-profit established in 2006 working to combat diseases through interventions to the built environment in vulnerable communities worldwide. She designs, implements, and evaluates projects that operate at the intersection of health and the built environment.

Her latest project focused on replacing dirt floors with concrete in Bangladesh to prevent diarrhoeal disease and respiratory and skin infections. Sarah has extensive experience in affordable housing, healthcare architecture, and construction, with a focus on sustainable design, resiliency, and hazard mitigation in vulnerable environments.

Jimi Patel, BDS, MPH, Research and Grants Officer, ARCHIVE Global
Jimi Patel graduated with her Master's degree in Public Health from Long Island University, Brooklyn in May 2017. She also holds a bachelor's degree in Dental Surgery from India. Jimi spent a summer in Belgrade, Serbia conducting quantitative and qualitative research on perceptions of health among young adults. She believes in health equity for all and is interested in epidemiology and global health. At ARCHIVE Global, she contributes to projects with her experience in grants, research, and evaluation.

Riksum Kazi, Interim Managing Director, ARCHIVE Global
Riksum Kazi served as the Interim Managing Director at ARCHIVE Global. He managed the organization's operations and projects around the world. Riksum graduated Phi Beta Kappa with dual degrees in International Political Economics and History and holds a Master's degree, with distinction, from Columbia University in Political Economics and Regional Affairs. Riksum has worked extensively in South Asia and is an active duty volunteer in the emergency services.

Charlotte Burch, Taubman College of Architecture and Urban Planning, University of Michigan
Charlotte Burch is a graduate from Pratt Institute with a Bachelor of Fine Arts in Interior Design and a minor in psychology. She joins ARCHIVE Global with a passion for designing positively impactful spaces to help those who need it the most. Throughout her studies, Charlotte has found herself studying in Copenhagen and working in London. She loves to learn more about the world and all the people in it. At ARCHIVE Global, Charlotte contributed to the research, design, and implementation of projects around the world.

Author's note

Sarah Ruel-Bergeron, Executive Director ARCHIVE Global; Jimi Patel, Research and Grants Officer, ARCHIVE Global; Riksum Kazi, Volunteer Consultant, ARCHIVE Global; Charlotte Burch, Project Officer, ARCHIVE Global.

Charlotte Burch is now pursuing dual master's degrees at the Taubman College of Architecture and Urban Planning at the University of Michigan. Riksum Kazi is now at Vital Strategies.

This needs assessment was supported in part by Dr Stephen Battersby and Joan Walley at the Campaign for the Renewal of Older Sewerage Systems (CROSS), UK.

Correspondence concerning this book should be addressed to Sarah Ruel-Bergeron, Executive Director, ARCHIVE Global, New York, NY 10003. Email: srb@archiveglobal.org

Routledge Focus on Environmental Health
Series Editor: Stephen Battersby, MBE PhD, FCIEH, FRSPH

Pioneers in Public Health
Lessons from History
Edited by Jill Stewart

Tackling Health Inequalities
Reinventing the Role of Environmental Health
Surindar Dhesi

Statutory Nuisance and Residential Property
Environmental Health Problems in Housing
Stephen Battersby and John Pointing

Housing, Health and Well-Being
Stephen Battersby, Véronique Ezratty and David Ormandy

Selective Licensing
The Basis for a Collaborative Approach to Addressing Health Inequalities
Paul Oatt

Assessing Public Health Needs in a Lower Middle Income Country
Sarah Ruel-Bergeron, Jimi Patel, Riksum Kazi, and Charlotte Burch

Assessing Public Health Needs in a Lower Middle Income Country

Sarah Ruel-Bergeron, Jimi Patel, Riksum Kazi, and Charlotte Burch

LONDON AND NEW YORK

First published 2021
by Routledge
2 Park Square, Milton Park, Abingdon, Oxon OX14 4RN

and by Routledge
52 Vanderbilt Avenue, New York, NY 10017

Routledge is an imprint of the Taylor & Francis Group, an informa business

British Library Cataloguing-in-Publication Data
A catalogue record for this book is available from the British Library

Library of Congress Cataloging-in-Publication Data
Names: Ruel-Burgeron, Sarah, author.
Title: Assessing public health needs in a lower middle income country / Sarah Ruel-Bergeron, RA, Director of Projects and Development, ARCHIVE Global.
Description: Milton Park, Abingdon, Oxon ; New York : Routledge, 2021. | Series: Routledge focus on environmental health | Includes bibliographical references and index.
Identifiers: LCCN 2020028802 (print) | LCCN 2020028803 (ebook) | ISBN 9780367530365 (hardback) | ISBN 9781003080220 (ebook)
Subjects: LCSH: Public health—Environmental aspects. | Public health—Methodology. | Sanitation.
Classification: LCC RA427 .R84 2021 (print) | LCC RA427 (ebook) | DDC 362.1—dc23
LC record available at https://lccn.loc.gov/2020028802
LC ebook record available at https://lccn.loc.gov/2020028803

ISBN: 978-0-367-53036-5 (hbk)
ISBN: 978-1-003-08022-0 (ebk)

Typeset in Times New Roman
by Apex CoVantage, LLC

Contents

Foreword

In 2018, ARCHIVE Global was commissioned to conduct a sanitation needs assessment for the municipality of Idenau, Cameroon with the generous support of Dr Stephen Battersby and Joan Walley of the Campaign for the Renewal of Older Sewerage Systems (CROSS). In close cooperation with Mr Rapheal Muma, Idenau Municipal Officer, ARCHIVE Global completed a sanitation needs assessment of Idenau in 2019. Together we are now seeking a path forward to implement a sanitation project for the community.

Idenau needs a comprehensive sanitation solution. With rampant open defecation, unaddressed industrial waste pollution in its drinking water sources, along with environmentally degrading activities, the residents of this municipality suffer from high rates of morbidity. This difficult situation calls for innovative solutions and systems to mitigate the challenges faced by the community. ARCHIVE and their partners have identified several solution pathways to create (1) more effective waste management systems from the household level to citywide infrastructure, (2) improvements to the water supply, and (3) a community-wide education campaign on adequate housing, sanitation, hygiene, and environmental protection. With the goal of creating lasting community-wide health improvements through a series of sanitation improvements, this book highlights structural challenges in this community and others like it and offers cost-effective strategies for change. The Participatory Rapid Appraisal method was used to develop a plan of action with the municipality to implement projects with other local and international partners. Our aim is to make our model replicable for similar communities affected by climate change, insufficient national funding, substandard infrastructure, inadequate housing, and associated health problems.

ARCHIVE Global is an award-winning international, 501(c)(3) registered non-profit organization that works at the intersection of health and the built environment to develop and implement cost-effective, scalable solutions in vulnerable communities worldwide. ARCHIVE is developing a multi-phase project to improve the water and sanitation infrastructure in the Idenau municipality in Cameroon with local and international partners. This effort is to be paired with an education campaign necessary to ensure the long-term success of creating an open defecation free municipality and improvements in the health of community members.

This book utilizes the United Nations' Sustainable Development Goals (SDGs) as the benchmark by which to measure Idenau's present status in achieving these outcomes. The authors recommend the municipality adopt the SDGs as a framework for benchmarking the social and economic development of the community. The book also outlines the priority SDG needs. Priorities include SDG 3: to ensure healthy lives and well-being for all ages; SDG 6: to ensure sustainable water and sanitation; SDG 11: to create an inclusive, safe, resilient, and sustainable city; and SDG 13: to combat climate change and its impacts. The implementation of these goals requires governmental cooperation and support, strengthened coordination, and effective mobilization and use of resources – public, private, local, and international.

The importance of conducting a needs assessment and disseminating results in communities like the Idenau municipality cannot be overstated. Compounding barriers make this type of investigation rare. ARCHIVE faced several of these challenges during this undertaking including a dearth of community and health data and lack of funding. Communicating with the local representative was another challenge faced due to poor telecommunication infrastructure. How Idenau officials, community-based organizations, underwriters, and outside funders tackle the issues raised in this book can set apart Idenau as an example of how to build a safer and cleaner community in resource limited settings.

See Appendix I for the letter from the Idenau municipality's local representative, Rapheal Muma, describing the current situation in Idenau and the value of the Needs Assessment, which forms the basis of this work.

Letter from Joan Walley

The following is a letter from Joan Walley describing the history of the relationship with Idenau and the value of the Needs Assessment, which forms the basis of this work.

Back in 2009 I visited Cameroon as a member of the House of Commons Environmental Audit Select Committee. The purpose was to further our inquiry into sustainably logged timber with a view to influencing UK government policy in respect of the outlawing of illegally logged timber, labelling of timber, introduction of robust timber procurement policy, and influencing of international agreements. In the course of that inquiry the High Commissioner arranged for the municipality of Idenau to host the Committee members, thereby offering the opportunity for us to see and understand first-hand the wider environmental challenges facing local people.

The warmth of that welcome has stayed with me. It was some years later that I was contacted by Rapheal. At the time of our visit he was a local student intent on completing his higher education and returning to his hometown to work to change things for the better, but by this time he was already working in Idenau and seeking support for the urgent sanitation problems and wider public health issues facing local people.

So I am delighted that The Campaign for the Renewal of Older Sewerage Systems (CROSS) in the UK, which is now closing its activities and which I chaired, has offered assistance to enable ARCHIVE Global to work alongside the municipality, Rapheal, and volunteers to devise and complete a Needs Assessment in Idenau. This is intended as a necessary first step in identifying the extent of recent local environmental challenges and providing an evidence base for their future action planning.

What this action follows might necessarily be determined by the detailed responses to the questionnaire.

One thing that is certain though is that the sanitation and environmental issues facing residents of Idenau are real. There is an increasing focus on the urgent need for implementation of the UN Sustainable Development Goals through partnership and collaboration. I very much hope that the initial survey work and volunteer involvement will therefore provide a springboard for local action and focus which can in a systematic and meaningful way lead to lasting changes in life expectancy and well-being of residents in Idenau and in the Commonwealth. I hope it will also bring in more partners and funding opportunities to secure these improvements.

~ Joan Walley

Acknowledgements

ARCHIVE Global would like to thank Rapheal Muma, the Honourable Tonde Lifanje Gabriel, Mayor of Idenau, and the volunteers of the Idenau municipality for all their hard work in assisting with data collection and information gathering for the Needs Assessment of Idenau. We would also like to acknowledge Dr Stephen Battersby, Joan Walley, and the Campaign for the Renewal of Older Sewerage Systems (CROSS) for their generosity and vision to help promote the use of the built environment as a powerful resource to improve health outcomes around the world. We would also like to provide special acknowledgement to key ARCHIVE Global members who provided critical support and work on the Needs Assessment, which forms the basis of this work. Special thank you to Araliya M. Senerat for significant contributions, including assisting with the literature review, designing the Cameroon Needs Assessment Survey, analysing both quantitative and qualitative data, assisting in writing the first draft, and editing the Needs Assessment; and Deanna Jiang for assisting with the literature review and the first draft of the Needs Assessment.

Acronyms and abbreviations

ARCHIVE Global:	Architecture for Health in Vulnerable Environments
CAMWATER:	Cameroon Water Utilities Corporation
CEMAC:	Central African Economic and Monetary Community
CDC:	Cameroon Development Corporation
CLTS:	Community-Led Total Sanitation
CROSS:	Campaign for the Renewal of Older Sewerage Systems
DALY:	Disability Adjusted Life Years
ESD:	Education for Sustainable Development
GESP:	Growth and Employment Strategy Paper
JMP:	The WHO/UNICEF Joint Monitoring Programme for Water Supply, Sanitation and Hygiene
MDG:	Millennium Development Goals
NGO:	Non-Governmental Organization
OD:	Open Defecation
ODF:	Open Defecation Free
OECD:	Organization for Economic Co-operation and Development
PNDP:	National Community-driven Development Programme
SDGs:	Sustainable Development Goals
UNFCCC:	United Nations Framework Convention on Climate Change
VNR:	Voluntary National Review
WASH:	Water, Sanitation, and Hygiene
WHO:	World Health Organization

Introduction

ARCHIVE Global believes that no one's health should be negatively impacted by the state of their housing. The built environment is a critical factor in building a healthy community. Household improvements by local tradesmen trained in innovative techniques, using locally available materials, promote the local economy and create healthier living environments for families. This cultivates more resilient communities that are better able to withstand crises.

ARCHIVE, Architecture for Health in Vulnerable Environments, works with vulnerable communities worldwide to improve living conditions and health outcomes through cost-effective, scalable built environment interventions, primarily focused on existing housing stock. ARCHIVE's work is based on a three-pronged approach of research, design, and advocacy.

1 Research: We collect baseline data with the participating families and conduct pre- and post-implementation household surveys. Monitoring and evaluating is done continuously, allowing us to iteratively improve each phase of an intervention based on community data and needs.
2 Design: Beginning with an exploration of the local cultural contexts and construction methods, we design and construct targeted and replicable interventions to existing homes and infrastructure within vulnerable communities.
3 Education and Advocacy: We deliver comprehensive, culturally appropriate and sensitive education campaigns to create lasting health improvements, specifically targeting women and children but with benefits to the entire family. We also focus on disseminating the message and lessons learned to a national audience, and to the public health and design communities worldwide to advocate for a better understanding of this critical intersection.

At our core, we use one basic right – housing – to deliver one basic need – health.

ARCHIVE has implemented multiple projects across the globe, from low-income countries to lower-middle and high-income countries. Examples of projects implemented by ARCHIVE are:

1. Building Malaria Prevention – Yaoundé, Cameroon: In 2012, Cameroon had 21,690,000 confirmed cases of malaria, and 71% of Cameroonians had malaria more than once. Eighty-five percent of settlements were informally or self-constructed, so screening and protection from malaria-bearing mosquitoes was mostly non-existent. To decrease the number of mosquitoes infiltrating the home and infecting the community, ARCHIVE directly worked to combat malaria through housing interventions in 264 homesteads in a resource-limited neighbourhood, using screened doors, windows, and eaves, along with adequate ventilation, sewage, and drainage solutions. These types of interventions can stop mosquitoes, which are vectors for transmitting diseases and mostly bite indoors, from entering the home, and thereby reducing the incidence of malaria. Over a span of three years, the project helped reduce indoor malaria-carrying mosquito exposure by 50% and malaria incidence by 20% for the participating families. Additionally, advocacy and education efforts afforded improvements for the entire community rather than only among the participants by enclosing wastewater pipes and clearing debris around the household which can pool water and breed mosquitoes. This intervention has since been replicated for ongoing projects in Namibia and Swaziland.
2. Mud to Mortar – Savar, Bangladesh: Vulnerable communities worldwide are subject to high risks of waterborne diseases linked to low-quality, substandard housing and sanitation infrastructure in addition to a lack of hygiene education and disease prevention. In Savar, Bangladesh, many villages suffer seasonal flooding, which brings contaminated water and liquid waste into roads, walking paths, and living environments. Compacted dirt flooring, extant in 67.8% of households in Bangladesh makes it easy for opportunistic parasites and bacteria to proliferate and cause life-threatening illnesses because the floors are nearly impossible to keep clean. Mud to Mortar delivers a unique intervention which helps address these challenges: the complete replacement of dirt floors with a simple concrete floor assembly combats diarrhoeal disease in the homes of vulnerable families. The project has, to date, completed nearly 300 floors, reducing the incidence of diarrhoeal disease by 52% among the participant families and proved that the flooring intervention has a number of health and social co-benefits, making the return on investment high.

3 Happy Healthy Households – Brent and Newham, London, UK: In 2010, London accounted for almost 40% of all TB cases in the UK with an increase in tuberculosis (TB) cases among immigrant neighbourhoods. The Happy Healthy Households project focused on increasing awareness of the relationship between poor health and poor housing. It also helped reduce disease stigmatization and incidences of tuberculosis among participants. Fifteen hundred community members were reached through engagement efforts, focus group discussions, and presentations. Participants were made aware of how environments that were damp, overcrowded, and lacked ventilation and adequate sanitation facilities could transmit tuberculosis. Three years post-intervention, the community saw a 30% reduction in TB hospitalizations through the efforts of many community-based campaigns.

Building on its expertise at the intersection of health and the built environment, ARCHIVE was commissioned to conduct a sanitation needs assessment for the municipality of Idenau, in the Southwest region of Cameroon. In 2018, ARCHIVE partnered with Idenau's Municipal Officer, Rapheal Muma, to conduct the initial assessment. By doing so, ARCHIVE discovered a vast data gap. The only available data source was the 2011 Idenau Council Development Plan, which was mandated by the government to benchmark and improve the living standards of Council areas.

A brief history of Idenau, Cameroon

Cameroon was formed in 1961 as an independent country following the unification of two former territories ruled by the British and the French. Cameroon is a culturally diverse Central African country, on the Gulf of Guinea, bordered by Chad, Nigeria, the Central African Republic (CAR), Gabon, Equatorial Guinea, and the Republic of the Congo. Cameroon's multiethnic community comprises about 250 groups that mainly fall under the three main groups: the Bantu, Semitic, and the Nilotic language groups. While 80% of the country's population are French speaking (Francophone), and 20% are English-speaking (Anglophone), the constitution of Cameroon promotes both English and French as the national languages of the country. The city of Yaoundé is the country's capital and the second largest city. Douala, the country's economic hub is the largest and the most populated city. See Figure 0.1.

Geographically, the country can be divided into four areas with 10 regions in total; northernmost, central, southern, and the western part. The Far North (Extreme Nord), North (Nord), and Adamawa (Adamaoua) are the three northernmost parts. The Centre and East (Est) are located centrally. The

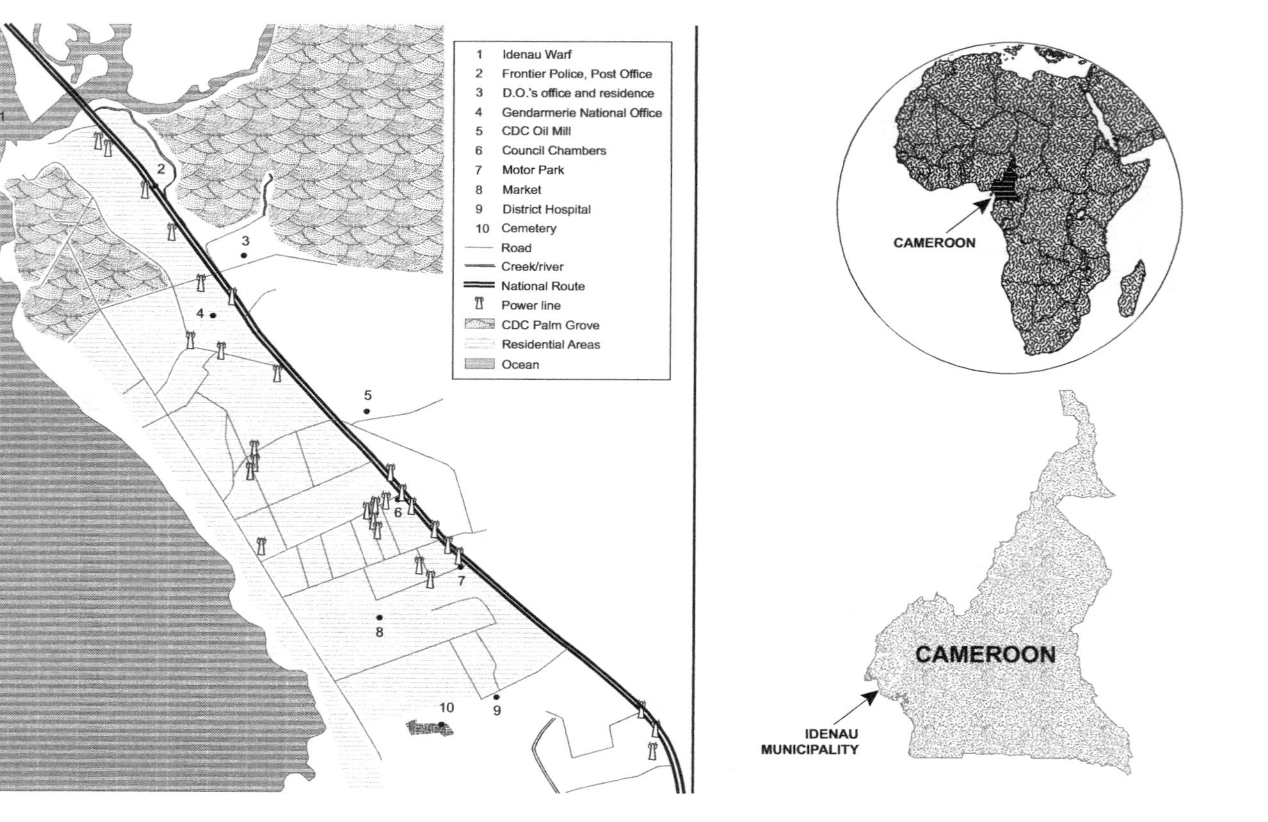

Figure 0.1 The geographical location of the Idenau municipality in the Southwest region of Cameroon

four regions that make the western part are the Littoral and the Southwest (Sud-Ouest) on the coast and the Northwest (Nord-Ouest) and West (Ouest) regions in the western grasslands. The South Province (Sud) region lies on the southern border and the Gulf of Guinea. The Northwest and the Southwest regions are Anglophone regions while the rest of the country is Francophone.

The Central African Economic and Monetary Community (CEMAC) comprises six countries – Gabon, Cameroon, CAR, Chad, the Republic of the Congo, and Equatorial Guinea – of which Cameroon has the largest economy. The diversified economy of Cameroon includes the export of oil and gas, timber, agriculture, and mining, of which oil accounted for nearly 40% of exports in 2019 despite the plunge in global oil prices. Similar to other lower middle-income countries, Cameroon's economy is affected by factors such as low per capita income, inequitable income distribution, and the internal conflict between the Francophone regions and the two Anglophone regions (Africa: Cameroon, n.d.). Adding to these tensions, Cameroon is affected by conflict in neighbouring nations, the threat of Boko Haram terrorists in the north, and civil unrest by groups seeking independence. The country also faces an influx of refugees from the neighbouring countries entering Cameroon which poses challenges for the distribution of resources and services.

The Idenau municipality, which is the subject of this publication, is located in the Anglophone (English-speaking) region, the Southwest region also known as Sud-Ouest region. The capital city of the Southwest region is Beau, located on the foot of Mount Cameroon. The city of Kumba is the largest city in the region and the city of Limbe, located 29 kilometres from the Idenau municipality, is a popular tourist destination for its beaches.

The Council of Idenau was created by the decree of the President on April 24, 1995. With a surface area of 16 kilometres, in the Southwest (Anglophone) region, the Idenau Council is Idenau municipality's principal town and is located 29 kilometres from the city of Limbe. The Idenau municipality is a conglomerate of villages that resides on the Atlantic coast of Cameroon and is an English-speaking community with an estimated population of 30,000 as of 2010.

Following the decentralization of powers to local councils on July 22, 2004, the Council, in partnership with two local organizations – Local Support Organization and Reach-Out Cameroon, led by the National Community-driven Development Programme (PNDP) – developed the Idenau Council Development Plan in 2011. The plan included the priorities and concerns of the municipality related to various sectors and industries, including agriculture and rural development; livestock, fisheries, and animal industries; forestry and wildlife; environment and nature protection; and energy and water resources (Idenau Council Development Plan, 2011).

The Idenau Council Plan (2011) identified some significant concerns within the municipality such as limited water and sanitation infrastructure. The issues regarding water and sanitation include dumping of household and other solid waste in and around the community, poor drainage systems with open gutters mixed with both sewage and rainwater, limited access to potable water for most residents, and poor road networks that further limit access for pumping trucks to empty septic tanks. Apart from the issues related to sanitation and waste management, limited town planning affecting infrastructure provisions, flooding, and environmental degradation such as deforestation are all issues of high concern to the residents of Idenau. Industrial waste such as sludge from oil mills is dumped in the streams, which pollutes the water and degrades the quality of the air, causing health concerns for the community.

ARCHIVE's research confirmed these challenges continue to exist and identified additional challenges faced by the community. Of all the issues identified, ARCHIVE focused on critical gaps in the municipality's infrastructure that affect water access, sanitation, waste management, and cause environmental degradation. Open defecation is a major health concern, as are diarrhoeal disease, malaria, and other conditions. Environmental degradation such as deforestation exacerbates issues like flooding, which increases dampness in households, creates visible mould and causes respiratory illnesses. The needs assessment maps out a series of proposed solutions to the highlighted water and sanitation challenges and provides several recommendations to improve conditions in the municipality.

Bibliography

Cameroon. (2020, April 2). *Encyclopaedia Britannica*. Encyclopaedia Britannica, Inc. Retrived from https://www.britannica.com/place/Cameroon

Cameroon: Confronting Boko Haram. (2016). Retrieved from Brussels, Belgium: www.crisisgroup.org/africa/central-africa/cameroon/cameroon-confronting-boko-haram

Idenau council development plan. (2011). Retrieved from PNDP, Southwest Regional Coordination Unit: www.pndp.org/documents/10_CDP_Idenau.pdf

Lange, K. (2019, July 25). *The current conflict situation in Cameroon*. Retrieved May 12, 2020, from https://ifair.eu/2019/07/25/the-current-conflict-situation-in-cameroon/

Ruppel, O. C., & Schlichting, K. R. (2018). Cameroon in a nutshell – Human and natural environment, historical overview and legal setup. In O. C. Ruppel & E. D. K. Yogo (Eds.), *Environmental law and policy in Cameroon – Towards making Africa the tree of life | Droit et politique de l'environnement au Cameroun – Afin de faire de l'Afrique l'arbre de vie* (pp. 51–74). Retrieved from www.jstor.org/stable/j.ctv941sr6.7

The World Factbook: Cameroon. (n.d.). Retrieved May 12, 2020, from www.cia.gov/library/publications/the-world-factbook/geos/print_cm.html

1 Background

The community

Idenau is a mid-sized municipality in the Sud-Ouest (Southwest) of the Republic of Cameroon. The municipality is located on the Atlantic coast along the windward side of Mount Cameroon. The average temperature ranges from 25°–30°C (77°–86°F) and the average monthly rainfall is about 5,000–8,000 millimeters (196–315 inches). The two distinct seasons are dry and rainy, with rainfall usually occurring from March to November. The last updated information available on the population size of the municipality was the 2010 census, which reported a totalof 30,000 residents living in eight villages including native communities, fishing ports, and Cameroon Development Corporation (CDC) camps. It is expected that the population of the municipality is now larger than 30,000 due to migration, increasing urbanization of the Council, and natural increase and population growth. The Cameroon Development Corporation (CDC) is an Agro-Industrial Corporation that operates all over Cameroon and grows, processes and markets tropical export crops such as bananas, semi-finished rubber, palm oil, and palm kernel oil. The CDC owns a major portion of the real estate in the municipality that is used for agriculture and palm oil mills. The CDC camps attract a considerable portion of skilled and unskilled workers from other regions in Cameroon. Apart from providing employment, the CDC also supports the municipality by maintaining the roads and supplying water to the camps and one other village, Njoni, due to its proximity to a CDC plantation (*Idenau Council Development Plan*, 2011).

The population of Idenau, Cameroon, includes a variety of ethnic groups, including Bakweri, Bamboko, and Bayange. Most of these groups are engaged in subsistence agriculture, fishing, and other income-generating activities including the civil service. Christianity is the predominant religion practiced by more than 90% of the population. Eighty percent of housing structures within Idenau are made of soft timber walls while the other 20% are made using concrete blocks and corrugated aluminium sheets. The fertile soil in the rural areas of the municipality attracts migrant farmers involved in the cultivation of major food crops such as plantains, cassava, cocoyam, and maize. Pidgin, English, and French are the dominant languages spoken in Idenau. Primary education is available for children in the municipality, but the quality of education is limited. School infrastructure, number of classrooms, equipment, and furniture for the students and teachers, potable water connections, toilets, and qualified teachers are lacking. The available nine health centres are poorly equipped and have insufficient personnel and infrastructure to ensure proper healthcare for the Idenau municipality (*Idenau Council Development Plan*, 2011).

Fishing is the principal economic activity in the municipality and is a major source of revenue for the council. Canoes and small-sized engine boats are used to carry out fishing in the community. This activity is mostly done by men while women are more involved in smoking and selling the fish. The transportation system is partially developed. Commercial motorbikes and vehicles are the main modes of transportation on the mainland area and boats are used for the maritime area.

The challenges

Health concerns

According to WHO (2019), each year, over 800,000 people worldwide die due to inadequate water, sanitation, and hygiene. Inadequate sanitation alone is responsible for an estimated 432,000 diarrhoea-related deaths annually; poor sanitation also contributes to several neglected tropical diseases, including intestinal worms, schistosomiasis, and trachoma. Open defecation causes tremendous health risks and diseases, including cholera, diarrhoea, dysentery, hepatitis A, typhoid, and polio. A combination of open defecation, malnutrition, and poverty is responsible for the highest number of deaths among children, globally.

The WHO/UNICEF Joint Monitoring Programme for Water Supply, Sanitation and Hygiene (JMP) developed a classification for the sanitation service ladder to compare the service levels across countries. The service level classification includes:

1 Open defecation, which is disposal of human faeces in open spaces.
2 Unimproved sanitation is the use of pit latrines without a slab or platform, hanging latrines, or bucket latrines.

3 Limited sanitation is the use of improved facilities shared between two or more households.
4 Basic sanitation means the use of improved facilities which are not shared with other households.
5 Safely managed sanitation is defined as the use of improved facilities which are not shared with other households where the excreta are safely disposed of.

In 2017, among the people in rural Cameroon, almost 14% practiced open defecation, which gradually increased from 13% in 2000 and is the highest value in the last 15 years. In 2017, only 39% of the total population of Cameroon had access to basic sanitation facilities, while 35% had unimproved sanitation, and 18.7% had limited sanitation. There are other risks in addition to infectious diseases caused by open defecation ("Household Data," 2019). In 2014, the UN raised the question of human safety and security with regard to dangerous animals and other threats. Girls especially are at risk of being kidnapped and raped when going outside in the fields to defecate after dark ("Ensuring women's access to safe toilets," 2014).

The consequences of unsafe water, sanitation, and hygiene (WASH) services can be grave in communities with low to no services. Unsanitary conditions are directly related to stunted growth, affecting vulnerable children under five years of age (Cumming & Cairncross, 2016). According to the WHO, the rate of mortality due to unsafe WASH services in 2016 was 45.2 per 100,000 people. Additionally, lack of sanitation infrastructure leads to environmental pollution due to uncontrolled exposure of waste, which contaminates clean water sources. Large pools of stagnant water act as mosquito breeding areas, increasing the risk of malaria. These and many other life-threatening problems are of concern in a community with unsafe WASH services.

Open defecation and lack of WASH infrastructure

Currently, a majority of the people in Idenau practice open defecation due to the lack of adequate toilets in the community. This refers to the percentage of the population defecating in the open, such as fields, forests, bushes, open bodies of water, on beaches, in other open spaces or disposed of as solid waste. Municipal infrastructure such as drainage and piped water is not currently available within Idenau. Limited access to potable water also remains a concern due to poor management of existing water sources and the breakdown of pumping engines, leading to the need to travel long distances to collect drinking water and a prevalence of water-borne diseases such as diarrhoea. According to the WHO, diarrhoea-related DALY in Cameroon in 2016 was 425,982, of which 258,652 are children under five.

Environmental concerns and waste management

Sludge from oil mills is dumped in the surface water bodies, which pollutes the water and degrades the quality of the air. The lack of proper waste management and a high rate of flooding are a few core problems in urban development and housing caused by haphazard dumping of refuse and poor drainage systems, respectively. This leads to environmental pollution, prevalence of airborne diseases, and increase in mosquito breeding areas.

The priority areas

The partnership between ARCHIVE and the municipality led to multiple conversations on the current situation in Idenau. The ARCHIVE partnership with Idenau revealed five recurring priorities for building and maintaining a healthy local community:

Potable water sources

Each year drinking contaminated water is the cause of death for over 500,000 individuals around the world ("Radical increase in water and sanitation investment," 2017). Similarly, the lack of access to clean potable water in the Idenau municipality has a direct effect on the health status of its inhabitants particularly affecting women, who are often responsible for collecting water for their families. According to UNICEF, women and girls spend an estimated 200 million hours every day collecting water, rather than attending school or participating in economically productive activities within the community ("UNICEF: Collecting water," 2016).

Access to clean water is a grave concern for the residents of the Idenau municipality. Although there are several freshwater sources throughout the municipality, they are polluted with industrial waste and contaminated by high levels of open defecation. Continuous access to potable water sources is limited requiring people to travel long distances, which limits the quantity that can be retrieved. Therefore, residents are inclined to store water for months in unsafe storage facilities like drums without lids. These practices and behaviours further lead to health issues including vector-borne diseases like malaria through mosquito breeding in stagnant pools of water and water-borne diseases like diarrhoea and typhoid from drinking contaminated water, specifically endangering the vulnerable groups of children and older adults in the communities. Health costs associated with water-borne diseases amount to more than a third of the income of poor households in Sub-Saharan Africa.

Hygiene and sanitation upgrades

The municipality of Idenau lacks the basic sanitation infrastructure needed for its inhabitants to follow proper hygiene behaviours. The lack of toilets in the community forces the residents to practice open defecation, introducing a cycle of faecal contamination and life-threatening infectious diseases. Hygiene behaviours like handwashing are vital in breaking the cycle of faecal contamination. However, the lack of access to clean water and absence of basic hand washing facilities prevent the residents of the Idenau municipality from learning or practicing handwashing. Historically, neither the residents nor the municipality have prioritized sanitation upgrades.

Apart from inadequate sanitation services, the municipality also lacks proper waste management services. Improper waste management is directly associated with numerous health risks, such as failure to collect and dispose of solid waste, which then leads to an increase in disease-carrying vectors like rodents and insects.

Environmental pollution

The Idenau municipality falls under maritime and equatorial forest areas, with coastal mangroves and equatorial rainforests. A vast majority of the biodiversity that existed in the forest has been depleted. Although a small portion of the forest is being maintained, a large portion of the forest has been cut down for the sole purpose of farming activities and timber exploitation. The rich diversity of flora and fauna that once existed in the area has been devastated due to unsustainable human exploitation. Timber is exploited for local and commercial uses, while animals like alligators, antelopes, cane rats, porcupines, squirrels, deer, and crocodiles have fallen prey to local hunting in the municipality. Activities like deforestation have been proven to increase the exposure to malaria and schistosomiasis in Africa.

There are various areas around Idenau with ecological importance. These areas, such as the sea, forest, mangroves, swamps, mountains, and creeks, are important for many reasons including their abundance in natural and mineral resources, for tourism and conservation, and yet do not have any environmental protections. These resources are in danger due to the irresponsible dumping of solid waste, human waste, and industrial waste. Incineration of household waste and indiscriminate dumping of human waste is quite common among the residents of the municipality. Waste from oil processing plants as well as sludge from the CDC oil mill near the water bodies is responsible for further polluting the environment. See Figure 1.1.

There are multiple fishing ports in the vicinity of the municipality, making fishing a major source of tax revenue collection for the Council. A high

Figure 1.1 From left to right: (1) image shows an example of oil sludge contaminating a water body in the Idenau municipality and (2) image shows a stream that runs through a neighborhood in Idenau. Like many other waterways found in the municipality, this body of water is used by the community to dispose of garbage, human excreta, and industrial by-products.

proportion of residents and migrants engage in fishing as an economic activity. Women in the local communities smoke fish and have to travel to markets in the nearby urban areas like Limbe city, 29 kilometres away, or sometimes all the way to Nigeria for a better market price. These activities – burning waste, deforestation for farming, smoking fish, and industrial production – have negative public health implications such as respiratory illness and contribute to environmental pollution (*Idenau Council Development Plan*, 2011).

Lack of infrastructure

The basic socio-economic infrastructure of Idenau includes two government health centres, two private dispensaries run by the CDC, three private clinics, two colleges of which one is a Government Technical College, a government high school, five public and three private nursery schools, nine public and four private primary schools, an electricity network and potable water outlets provided by the CDC, sectoral offices, CDC plantations, and two banks (*Idenau Council Development Plan*, 2011).

In addition to fishing, agriculture is the second main source of income, most of which is done on land owned by the CDC. A variety of food is cultivated on the CDC plantations, such as plantains, cassava, cocoyam, egusi, maize, oil palms, cocoa, as well as vegetables, and coconuts. Some livestock and animal rearing is also practiced. Animal rearing involves primarily pigs, goats, fowls, dogs, and cows (*Idenau Council Development Plan*, 2011).

Most of these animals are stray and may act as carriers of parasites like ticks, fleas, mites, Ascaris, Strongyles, and tapeworm.

The current overall infrastructure provisions of Idenau are not well developed. Housing is haphazard, with more than 90% of the homes being self-built. As such, structural integrity and the choice of material used when building an adequate home is not likely to promote health. Very few schools in Idenau are built as permanent structures while the rest of the schools have semi-permanent structures with no latrines and inadequate classrooms and desks. The roads that connect the residents to the schools are built poorly with temporary bridges, making it difficult and sometimes dangerous for the children. Lacking reliable access to electricity is a common problem, caused by either no connection to the rural electricity network or by inadequate (low voltage) electricity supply mechanisms in some villages. The majority of the houses are built using found materials for the walls that are inappropriate for construction, earthen floors, and corrugated aluminium sheets for the roofs. Self-built homes are built in an urgency to create a shelter from the environment, and often the materials that are readily available or affordable in these conditions are even scarce. Materials that are not intended for the purpose of building homes are inherently flawed. For example, plastic sheeting does not offer protection besides a visual barrier and corrugated aluminium sheets exacerbate indoor temperatures causing heat-related illness and non-communicable health conditions. Houses with dirt floors host disease-carrying organisms; disease transmission is rampant in small and overcrowded homes; access to running water and sanitation facilities in these conditions is essentially non-existent, causing WASH-related illnesses. The average home has three to five rooms with an average occupancy of two people per room (*Idenau Council Development Plan*, 2011). Due to the rapid increase in the population caused mainly by fishing, agriculture, and the illegal timber transportation industry, most of the residential areas are informal settlements that were unplanned and have grown into either a linear or clustered pattern without proper access to basic facilities like water and sanitation.

Education campaigns

A gap in knowledge exists on the importance of and strategies for healthy sanitation and hygiene practices. The lack of awareness of proper hygiene behaviours and sanitation practices is a direct risk factor for infectious diseases and hygiene-related health issues such as cholera, yellow fever, chronic diarrhoea, trachoma, scabies. There is also a need for awareness among the residents of the Idenau municipality on the short- and long-term impact of their environmentally degrading activities like deforestation for the purpose of farming, oil spills, exploitation of timber, and hunting and poaching of forest animals. While planning an education campaign, it is important to use culturally and socially appropriate messaging that is relatable to a specific community. Based on ARCHIVE's experiences, culturally sensitive knowledge retention exercises have proven to be very effective among our participants in the Mud to Mortar project in Bangladesh.

Community-led Total Sanitation (CLTS) is an approach to eliminating open defecation (OD) in communities. This is a marked shift from the focus on providing individual households with sanitation infrastructure to focus instead on community-level education about the impacts of open defecation. Apart from the community level education, there is a need to train the municipality authorities in soft skills, such as advocacy and supervision of the community to ensure a successful CLTS campaign.

This publication describes suggested changes that focus mainly on water, sanitation, and hygiene concerns of Idenau primarily, although measures to combat these problems also address other identified concerns like waste management and environmental pollution. The local economy can be improved by developing a financially sustainable sanitation service chain. The education component addresses health literacy through improvements in hygiene and sanitation behaviours and good environmental practices. Improved health literacy has many long-term benefits, including improvements in health conditions and reductions in the number of work and or school days missed.

Bibliography

Collecting water is often a colossal waste of time for women and girls. (2016, August 29). Retrieved May 12, 2020, from www.unicef.org/press-releases/unicef-collecting-water-often-colossal-waste-time-women-and-girls

Coluzzi, M. (1984). *Heterogeneities of the malaria vectorial system in tropical Africa and their significance in malaria epidemiology and control*. Retrieved May 14, 2020, from www.ncbi.nlm.nih.gov/pmc/articles/PMC2536202/

Cumming, O., & Cairncross, S. (2016). Can water, sanitation and hygiene help eliminate stunting? Current evidence and policy implications. *Maternal & Child Nutrition, 12*, 91–105. https://doi.org/10.1111/mcn.12258

Ensuring women's access to safe toilets is "moral" imperative, says Ban marking World Day | | UN News. (2014, November 19). Retrieved May 12, 2020, from https://news.un.org/en/story/2014/11/484042-ensuring-womens-access-safe-toilets-moral-imperative-says-ban-marking-world-day

GHO | World Health Statistics data visualizations dashboard | Unsafe water, sanitation, hygiene, (WASH) services. (2019). Retrieved May 12, 2020, from https://apps.who.int/gho/data/node.sdg.3-9-viz-2?lang=en

Guerra, C. A., Snow, R. W., & Hay, S. I. (2006, April). *A global assessment of closed forests, deforestation and malaria risk*. Retrieved May 14, 2020, from www.ncbi.nlm.nih.gov/pubmed/16630376

Household data. (n.d.). Retrieved May 12, 2020, from https://washdata.org/data/household#!/table?geo0=country&geo1=CMR

Idenau council development plan. (2011). Retrieved from www.pndp.org/documents/10_CDP_Idenau.pdf

Public Health and Environment (PHE): Water, sanitation and hygiene attributable burden of disease (low- and middle-income countries), 2016 Inadequate water: Diarrhoea DALYs in Children under five years. (2018). Retrieved May 12, 2020, from http://gamapserver.who.int/gho/interactive_charts/phe/wsh_mbd/atlas.html

Radical increase in water and sanitation investment required to meet development targets. (2017, April 13). Retrieved May 12, 2020, from www.who.int/en/news-room/detail/13-04-2017-radical-increase-in-water-and-sanitation-investment-required-to-meet-development-targets

Sanitation. (2019, June 14). Retrieved May 12, 2020, from www.who.int/news-room/fact-sheets/detail/sanitation

Taylor, D. (1997, November). *Seeing the forests for the more than the trees*. Retrieved May 14, 2020, from www.ncbi.nlm.nih.gov/pmc/articles/PMC1470324/

Yasuoka, J., & Levins, R. (2007). *Impact of deforestation and agricultural development on anopheline ecology and malaria epidemiology*. Retrieved May 14, 2020, from www.ajtmh.org/docserver/fulltext/14761645/76/3/0760450.pdf?expires=1589484778&id=id&accname=guest&checksum=2349D8FDE7ED20A59FAE88195716245E

2 Data collection methodology

ARCHIVE was commissioned by CROSS UK to conduct a Sanitation Infrastructure Needs Assessment for Idenau. ARCHIVE partnered with the Idenau municipality thereafter to produce the needs assessment that forms the basis for this book. Ultimately, this book is intended to highlight the need to create viable strategies to improve sanitation infrastructure for the community in the municipality. It will also be the basis of a road map to outline the sequence of work packages or actions that will need to be implemented to improve the community's infrastructure and will serve to seek funding in a series of phases. ARCHIVE aims to deploy specific measures in collaboration with the municipality and other relevant partners from various sectors.

The needs assessment objectives

Overall objective

The objective is to identify, assess, and analyze challenges and opportunities in the area of disease transmission, and prevention, and the barriers present in promoting health within the existing built environment. This includes outlining appropriate actions that are tailored to this community's specific needs around water and sanitation infrastructure, and education campaigns to support its development.

Objective 1. To conduct a needs assessment for Idenau, Cameroon, to understand the community's WASH infrastructure, and waste management customs.

Objective 2. To propose infrastructure recommendations to improve residents' health and hygiene.

Objective 3. To implement adequate sanitation, waste management, and potable water infrastructure within the Idenau municipality.

ARCHIVE initiated the needs assessment by focusing on understanding the concerns of the residents in Idenau and defining the purpose of the needs assessment. Starting with the secondary data collection, which refers to the readily available data or previously collected data, we realized early on during this stage that there was insufficient literature and health data available. There was also a gap in updated, reliable data such as demographic data, health data, and geographic data of the Idenau municipality. Support from a local representative from the Idenau municipality provided the basic information on the municipality, and enabled ARCHIVE to characterize the population.

The needs assessment includes a secondary data collection, which is a literature review of existing information, to identify the current situation and ongoing interventions on WASH followed by a literature review of evidence-based interventions implemented in similar communities. The primary data collection (quantitative and qualitative data) was conducted through surveys and a community discussion of needs, challenges, and opportunities on existing water and sanitation infrastructure, household waste management activities, and health issues. Post primary data collection, the analysis was conducted on how the sanitation infrastructure goals would fit within the context of the other priorities outlined by the municipality. The needs assessment also included an outline of suggested strategies for development and opportunities and challenges associated with implementing interventions in the Idenau municipality.

Methodology

Secondary data collection

The preliminary stage of the Needs Assessment consisted of secondary data collection exploring the current sanitation infrastructure conditions in Idenau. Prior to the start of the literature review, the Idenau municipality provided ARCHIVE with documentation on the environment, culture, lifestyle, demographic, and economic makeup of those living in Idenau, Cameroon. Phone call discussions were held with a representative from the Idenau municipality to discuss any questions or concerns regarding the documentation. The representative provided ARCHIVE with information regarding the current sanitation infrastructure and the common steps for waste disposal.

After the authors learned more about the population and infrastructure of Idenau, a thorough literature review was conducted on Cameroon, other African countries, and around the world. A literature review was also conducted to examine different surveys conducted in places similar to the population of the Idenau municipality and environment, to examine successful and unsuccessful interventions of sanitation infrastructure and review common behaviours of the residents of Idenau.

Primary data collection

Primary data collection was conducted using a mixed method approach including quantitative and qualitative data collection, which followed the Participatory Rapid Appraisal method to conduct the needs assessment. Participatory Rapid Appraisal is an informal and semi-structured method of data collection at the community level which ensures a high level of participation of the target group. Volunteers from the municipality were trained to conduct surveys and a community discussion to identify, assess, and analyze challenges and opportunities in the area of disease transmission and prevention, local conditions, and traditional approaches to design and construction.

Based on the literature review, questions from five surveys were used to create the needs assessment survey conducted in Idenau. The questions included in the survey for the residents were based on their sanitation behaviour, their access to potable water, health concerns, and their demographic data. The survey was reviewed by the authors and a representative from the Idenau municipality before sending it to the mayor of Idenau for deployment of volunteers into the community.

Quantitative data collection

Researchers targeted 1,000 residents in Idenau to complete the survey, interviewing a resident per household of age 18 and above. Volunteers from the municipality were trained on how to conduct the survey with the residents and explained how to gain consent for the survey. Consent was given by participants via a signature or fingerprint. Volunteers asked residents survey questions and completed the survey by hand. The survey was written and spoken in English. If the residents did not speak English, the questions were translated and asked in Pidgin, a common language in Idenau. The needs assessment survey is included in Appendix II.

Qualitative data collection

A community discussion with 42 participants and two one-on-one interviews was conducted among the residents of Idenau. Consent for recording the discussion was taken via signature or fingerprint from the respondents. The discussion was formulated to facilitate a conversation among the local representatives of the community regarding problems in the community, beliefs regarding health issues, and sanitation infrastructure within Idenau.

Questions and the consent form for the community discussion can be found in Appendices III and IV, respectively.

Quantitative and qualitative data analysis

The needs assessment was reviewed and approved as a Not Human Subjects Research Under 45 CFR 46 by Columbia University's Research Administration System. Data from the survey were collected on paper and later digitized in Microsoft Excel. The identifiers of all respondents were removed, and a unique identification number was allotted to each respondent before transferring the data into a computer to protect the identities of the participants. The representative from the Idenau municipality sent all data to ARCHIVE for data analysis. Data were cleaned in Excel and analyzed using descriptive statistics. Responses from open-ended questions were grouped based on themes in Microsoft Word, and descriptive statistics were used to calculate the percentages of common responses.

Bibliography

Core questions on drinking water and sanitation for household surveys. (2006). Retrieved May 14, 2020, from www.who.int/water_sanitation_health/monitoring/oms_brochure_core_questionsfinal24608.pdf789241563260_eng.pdf;sequence=1

Core questions and indicators for monitoring WASH in Schools in the Sustainable Development Goals. (2016). Retrieved May 14, 2020, from https://washdata.org/sites/default/files/documents/reports/2018-08/SDGs-monitoring-wash-in-schools-2018-August-web2.pdf

Davis, J., Garvey, G., & Wood, M. (1993). *Developing and managing community water supplies*. Oxford: Oxfam.

Enquête Démographique et de Santé et à Indicateurs Multiples 2011. (2011). Retrieved May 14, 2020, from http://microdata.worldbank.org/index.php/catalog/1564

Rijsdijk, A., & Mkwambisi, D. (2016, June). *Evaluation of the water and sanitation (Wash) programme in* . . . Retrieved May 14, 2020, from www.unicef.org/evaldatabase/files/Evaluation_of__Malawi_WASH_Programme_Malawi_2016-001.pdf

3 Literature review

Sixty-two articles were reviewed for the purpose of conducting a literature review. Forty-eight articles focused on water, hygiene, sanitation, and waste management. The other articles focused on open defecation and disease transmission. Thirty of the articles were focused in Cameroon, 15 from other countries in Africa, and two from Bangladesh. The rest of the articles were not location specific. The keywords used for the literature review were Cameroon sanitation, Cameroon sanitation infrastructure, Cameroon open defecation, Cameroon WASH, Senegal sanitation and faecal management infrastructure, Community-Led Total Sanitation, and waste management infrastructure.

Access to safe drinking water

In countries around the world, sanitation and access to drinking water plays a major role in the health of children and families. The common sources of drinking water in Cameroon include piped water distributed by the national water company, CAMWATER, for a monthly subscription fee, boreholes, wells, communal standpipes, and springs.

There is abundant research showing the positive impacts of access to clean water and sanitation on economic development and health. Studies conducted by Sanou et al. (2015) and Kuitcha, Kabeyene, Nkamjou, Lienou, and Ekodeck (2008) showed that the lack of sanitation and water infrastructure plays a large role in water quality and sanitation conditions. Access to safe water and sanitation is important for development and poverty reduction – factors that are necessary for good health and well-being (Alagidede & Alagidede, 2016). A study conducted by Ako, Shimada, Eyong, and Fantong (2010) on the access of potable water in Cameroon found the availability of water in Cameroon was not an issue, but access to safe drinking water was the most concerning issue. Most households who cannot afford to pay for piped water depend on other groundwater sources like wells, communal standpipes, etc. Kuitcha et al. (2008) in their study to analyze the water and sanitation situation in Yaoundé, Cameroon, questioned the quality of the groundwater in communities with ill-planned settlements due to the proximity of the groundwater to pit latrines, which are commonly used within communities with informal settlements. The study conducted by Sanou et al. (2015) in Douala, Cameroon, with a sample of 285 households from 14 districts found that 52.5% of the public water points were non-functional and 83% of the well water analyzed for water quality was positive for faecal contamination. The authors also found the age group that was mostly affected by water-borne diseases like cholera were the children under the age of five at 43%.

The impact of urbanization and migration to cities has led to a decrease in access to drinking water for the urban population of Cameroon, with a decrease from 85% in 2000 to 77% in 2017. On the other hand, in rural Cameroon, access to basic drinking water in 2000 was 38%, which increased by 1% through 2017 ("Household Data" 2019). The increase in urban population causes an overspill of the population into peri-urban areas, with a mix of rural and urban poor who cannot afford the cost of living in the city. Fonjong and Fokum (2017) conducted a study to examine the effects of privatization of the water sector on the water crisis in Cameroon and found no improvement in the water problems of the peri-urban communities. The authors explained the importance of urban planning and the impact of the sudden emergence of peri-urban areas, which are more likely to be self-constructed and ill planned, making the lives of the families living here more challenging with the lack of access to basic facilities like safe water and sanitation.

Open defecation and lack of WASH infrastructure

Open defecation is a common problem in Cameroon and leads to worse hygiene conditions. A study conducted by Ngwa et al. (2017) explored cultural practices and beliefs influencing cholera transmission in Cameroon and found that residents recognized the risks of open defecation and poor food and water quality as factors influencing the transmission of cholera. However, respondents could not culturally or financially avoid them. Some causes for the spread of cholera included funeral practices, which expose attendees to airborne diseases when eating at gatherings, communal water usage, groups eating from the same bowl, and public health messages that were written in a language not understood by the target audience. There was also a belief that too much consumption of pharmaceutical products destroys human organs and traditional medicine is superior. Some respondents also believed cholera was punishment from the gods. The lack of sanitation and water infrastructure plays a large role in health risks of Cameroon residents, but the study also showed the importance of considering the local beliefs, common practices, and behaviours to promote health among individuals.

Some research has been conducted on improving WASH infrastructure in vulnerable communities. In both Bangladesh and Kenya, a cluster-randomized controlled study combined with WASH and nutrition intervention was done to determine its effects on early child development for children under three years of age, including diarrhoea, and growth faltering, at one-year and two-year follow-ups post-intervention. A sample of 5,551 pregnant women in their first or second trimester were randomly selected from eight geographically adjacent clusters and randomly assigned to six intervention groups. The six interventions were chlorinating; improved sanitation; handwashing with soap; combined water, sanitation and handwashing; improving nutrition through counselling along with lipid-based nutrient supplements; and combined water, sanitation, handwashing, and nutrition. These studies conducted by Luby et al. (2018), Null et al. (2018), Stewart et al. (2018), Tofail et al. (2018) compared all interventions separately and together and found that children's diarrhoea prevalence in Bangladesh decreased in households with all interventions except water treatment but did not decrease in Kenya. Null et al. suggested this could be due to the possibility that the water, sanitation, and handwashing intervention did not sufficiently address important transmission routes for enteric pathogens. Nevertheless, there were small, significant improvements in children's motor development in WASH and nutrition interventions.

Another study conducted by Butala, VanRooyen, and Patel (2010) in Ahmedabad, India, demonstrated statistically significant results of a causal relationship between basic infrastructural improvement in water supply and sanitation and a reduction in the incidence of water-borne illnesses. The improvement strategy included a set of interventions: connection to water supply, toilet, and underground sewage for each household; storm water drainage, stone paving of internal and approach roads, solid waste management, and street lighting. Apart from the positive health outcomes, infrastructure improvements also have a direct effect on the social determinants of mental health, improving the safety of women, who otherwise are usually responsible for traveling long distances to collect water, allowing women to miss fewer days of work due to illness, and the decreased hospital and medical costs, all while increasing overall productivity and wealth.

Community-led total sanitation

Historically, sanitation interventions that have focused solely on providing infrastructural subsidies to households have only partially solved sanitation issues in communities (Chambers, 2009). One approach that has proven its efficacy is the Community Led Total Sanitation (CLTS). CLTS was pioneered by Kamal Kar, a development consultant, in 2000, in a village in the Rajshahi district of Bangladesh while evaluating a top-down subsidised sanitation programme. CLTS is an approach that shifts the focus from an individual or household level to a community-wide behavioural change to eliminate open defecation in communities. Communities are encouraged to conduct their own assessment and analysis of open defecation and are encouraged to take action themselves to become open defecation free (The CLTS approach, 2018). A successful CLTS intervention comprises three stages: (1) pre-triggering stage in which a community is selected, facilitators are trained by the CLTS implementing NGO or local partners, baseline community data are collected, and a strategy is developed to build rapport and coordinate an entry into the community; (2) triggering stage in which a community meeting is organized where facilitators conduct participatory exercises to stimulate a sense of shame and disgust to motivate the community members to take their own actions and change the sanitation situation; and (3) post-triggering stage in which routine monitoring and follow-up visits are conducted by the facilitators with the goal of verifying and certifying Open Defecation Free (ODF) status in community (Kar & Chambers, 2008). There are certain key principles for the CLTS intervention, which includes the shift from the standard top-down sanitation or latrine design by the facilitators, no individual household subsidies, allowing the community to create the plan that is most effective for themselves through facilitation and not teaching or preaching, and to install toilets in the post-triggering stage using existing resources in the community (Chambers, 2009).

UNICEF and the UK Department of International Development supported a CLTS programme for 600 residents in the village called Tilorma, Sierra Leone, in partnership with the Ministry of Health and Sanitation, a local NGO, and community organizations. As a part of the programme, natural leaders were identified from the community during pre-triggering activities like transect walks to understand how flies transmit the excreta and how open defecation causes dangerous health consequences within the community. This triggered the community to install 30 pit latrines for 600 villagers using available resources from the community. The community was certified ODF after six months following the introduction of CLTS. A cluster-randomized controlled study was conducted by Pickering, Djebbari, Lopez, Coulibaly, and Alzua (2015) in Mali, to assess the impact of CLTS programme among 60 villages in the intervention group compared to 61 villages in the control group. The CLTS programme was carried out in three stages. Every two to four weeks, water samples and health data were collected from each village.

The health-related data measured were diarrhoea, height for age, weight for age, stunting, and underweight. The results showed no difference in diarrhoeal prevalence among the children between the two groups. However, the children in the villages that received the CLTS intervention were taller and less likely to be stunted or underweight when compared to those from the villages in the control group.

Lack of domestic waste management

Apart from lack of water and sanitation infrastructure, another considerable challenge for the residents of Idenau is the lack of solid waste management domestically and commercially. Poor waste management is a risk factor for vector-borne and infectious diseases (Krystosik et al., 2019). Unplanned and haphazard dumping of household and human waste in the water bodies and around the community poses a great risk to the health of the residents in Idenau, their environment, and the marine life in the ocean (*Idenau Council Development Plan*, 2011). Waste management education and strategies are vital for a healthy future for the next generation and for today. There is an urgent need for more sustainable, cost-effective ways of waste management through systems and facilities.

There are successful and cost-effective waste management strategies being implemented in African countries such as Kenya, Uganda, and Ghana. The Clean Team in Ghana is a social enterprise providing toilet and regular waste collection services to households with five individuals for an affordable monthly fee of approximately $7 a month. The Clean Team collects the waste from these households and safely transports it to transfer stations via tuk tuks where it is then further transported to a central facility by trailers or tractors to be disposed of either into a septic tank, a dumping site, or transferred to a waste treatment facility. The sanitation process used by Sanergy in Nairobi, Kenya, costs about $14 per person per year, as opposed to the cost for the government to provide sanitation for a person per year at $54. Sanergy provides low-cost, high-quality toilet units designed to separate liquid and solid waste at the source for easy collection and treatment of the waste.

A team from Water For People in Kampala, Uganda, implemented a successful sanitation model by supporting the local community through affordable options for toilets by inventing new types of toilets and safe waste management techniques, testing new sanitation approaches in local markets, and scaling sanitation service strategies in partnership with the community members. Their model also provides piped water and utilizes the private sector to manage the water systems and create jobs in the community for mechanics and water kiosks and hand pump attendants.

Lack of non-domestic waste management

The industrial waste from the CDC oil mill and other smallholder oil processing units located near the water bodies in Idenau is an additional concern for the municipality. The global demand for palm oil for food, industrial transformation, and biofuel production has led to the conversion of vast lands of natural forests to palm oil plantations. The rapid expansion of the palm oil industry is a major cause of the destruction of ecosystems through deforestation, loss of biodiversity, and pollution (Kome & Tabi, 2020). The oil spills from smallholder oil processing units and sludge from the CDC oil mill contaminate the water and air. Residents of Idenau depend on these water bodies, including the ocean, local streams, and rivers for fishing, and use freshwater bodies as a source of drinking water for themselves and their livestock (*Idenau Council Development Plan*, 2011). The marine life in Cameroon is one of the main lifelines for economic activities and is at risk of being depleted by pollution and over-exploitation.

Bibliography

Ako, A. A., Shimada, J., Eyong, G. E. T., & Fantong, W. Y. (2010). Access to potable water and sanitation in Cameroon within the context of Millennium Development Goals (MDGS). *Water Science and Technology, 61*(5), 1317–1339. https://doi.org/10.2166/wst.2010.836

Alagidede, P., & Alagidede, A. N. (2016). The public health effects of water and sanitation in selected West African countries. *Public Health, 130*, 59–63. https://doi.org/10.1016/j.puhe.2015.07.037

Butala, N. M., VanRooyen, M. J., & Patel, R. B. (2010). Improved health outcomes in urban slums through infrastructure upgrading. *Social Science & Medicine, 71*(5), 935–940. https://doi.org/10.1016/j.socscimed.2010.05.037

Chambers, R. (2009). Going to scale with community-led total sanitation: Reflections on experience, issues and ways forward. *IDS Practice Papers, 2009*(1), 1–50. https://doi.org/10.1111/j.2040-0225.2009.00001_2.x

Fonjong, L., & Fokum, V. (2017). Water crisis and options for effective water provision in urban and peri-urban areas in Cameroon. *Society & Natural Resources, 30*(4), 488–505. https://doi.org/10.1080/08941920.2016.1273414

Household data. (2019). Retrieved from https://washdata.org/data/household#!/table?geo0=country&geo1=CMR

Idenau council development plan. (2011). Retrieved from PNDP, Southwest Regional Coordination Unit: www.pndp.org/documents/10_CDP_Idenau.pdf

Kar, K., & Chambers, R. (2008). *Handbook on community-led total sanitation.* Retrieved from www.communityledtotalsanitation.org/sites/communityledtotalsanitation.org/files/media/cltshandbook.pdf

Kome, G. K., & Tabi, F. O. (2020). Towards sustainable oil palm plantation management: Effects of plantation age and soil parent material. *Agricultural Sciences, 11*(1), 54–70. https://doi.org/10.4236/as.2020.111004

Krystosik, A., Njoroge, G., Odhiambo, L., Forsyth, J. E., Mutuku, F., & LaBeaud, A. D. (2019). Solid wastes provide breeding sites, burrows, and food for biological disease vectors, and urban zoonotic reservoirs: A call to action for solutions-based research. *Front Public Health*, *7*, 405. https://doi.org/10.3389/fpubh.2019.00405

Kuitcha, D., Kabeyene, B. V. R. K., Nkamjou, L. S., Lienou, G., & Ekodeck, G. E. (2008). Water supply, sanitation and health risks in Yaounde, Cameroon. *African Journal of Environmental Science and Technology*, *2*(11), 379–386. Retrieved from www.academicjournals.org/AJest

Luby, S. P., Rahman, M., Arnold, B. F., Unicomb, L., Ashraf, S., Winch, P. J., . . . Colford, J. M. (2018). Effects of water quality, sanitation, handwashing, and nutritional interventions on diarrhoea and child growth in rural Bangladesh: A cluster randomised controlled trial. *The Lancet Global Health*, *6*(3), e302–e315. https://doi.org/10.1016/s2214-109x(17)30490-4

Ndjama, J. P., Beyala, V. R. K. K., Nkamdjou, L. S., Ekodeck, G., & Awah, T. M. (2008). Water supply, sanitation and health risks in Douala, Cameroon. *African Journal of Environmental Science and Technology*, *2*(12), 422–429. Retrieved from www.academicjournals.org/AJEST

Ngwa, M. C., Young, A., Liang, S., Blackburn, J., Mouhaman, A., & Morris, J. G., Jr. (2017). Cultural influences behind cholera transmission in the Far North Region, Republic of Cameroon: A field experience and implications for operational level planning of interventions. *The Pan African Medical Journal*, *28*, 311. https://doi.org/10.11604/pamj.2017.28.311.13860

Null, C., Stewart, C. P., Pickering, A. J., Dentz, H. N., Arnold, B. F., Arnold, C. D., . . . Colford, J. M. (2018). Effects of water quality, sanitation, handwashing, and nutritional interventions on diarrhoea and child growth in rural Kenya: A cluster-randomised controlled trial. *The Lancet Global Health*, *6*(3), e316–e329. https://doi.org/10.1016/s2214-109x(18)30005-6

Pickering, A. J., Djebbari, H., Lopez, C., Coulibaly, M., & Alzua, M. L. (2015). Effect of a community-led sanitation intervention on child diarrhoea and child growth in rural Mali: A cluster-randomised controlled trial. *The Lancet Global Health*, *3*(11), e701–e711. https://doi.org/10.1016/s2214-109x(15)00144-8

Sanou, M. S., Temgoua, E., Guetiya, R.-W., Alyexandra, A., Losito, F., Fokam, J., . . . Vittorio, C. (2015). Water supply, sanitation and health risks in Douala 5 municipality, Cameroon. *Igiene e Sanità Pubblica*, *71*, 21–37.

Soap stories and toilet tales: 10 case studies. (2009). Retrieved from www.unicef.org/wash/files/8_case_study_SIERRA_LEONE_4web.pdf

Stewart, C. P., Kariger, P., Fernald, L., Pickering, A. J., Arnold, C. D., Arnold, B. F., . . . Null, C. (2018). Effects of water quality, sanitation, handwashing, and nutritional interventions on child development in rural Kenya (WASH Benefits Kenya): A cluster-randomised controlled trial. *The Lancet Child & Adolescent Health*, *2*(4), 269–280. https://doi.org/10.1016/s2352-4642(18)30025-7

The CLTS approach. (2018, February 14). Retrieved from https://www.communityledtotalsanitation.org/page/clts-approach

Tofail, F., Fernald, L. C. H., Das, K. K., Rahman, M., Ahmed, T., Jannat, K. K., . . . Luby, S. P. (2018). Effect of water quality, sanitation, hand washing, and nutritional interventions on child development in rural Bangladesh (WASH Benefits Bangladesh): A cluster-randomised controlled trial. *The Lancet Child & Adolescent Health*, *2*(4), 255–268. https://doi.org/10.1016/s2352-4642(18)30031-2

4 Quantitative data analysis results

Current demographics

A total of 988 participants took part in the survey, Table 4.1 shows the demographic information of the respondents of the needs assessment survey. Of the participants, 57.8% were women and 42.1% were men. When interviewing participants, only 37.8% identified themselves as head of the household. Of those who weren't heads of the household, about 60% of the participants identified as the spouse of the head of the household. The majority of the participants were between 35 and 44 years old (36.3%), followed by 25–34 years old (30.9%), 45–54 years old (27.3%), 18–24 years old (2.8%), and 55–64 years old (2.5%). Forty-nine percent were of Grassfields ethnicity, followed by Bamilike/Bamoun (18.7%) and Bantoïde Sud-Ouest (14.6%). The majority of participants were Catholic (43%) and Protestant (33.9%), with a number of participants identifying as having no religion (19.3%). When asked about the highest level of education, 13.9% completed superior (age 19–23 years), 17.7% completed secondary second cycle (age 17–19 years), 29.6% completed secondary 1st cycle (age 12–17 years), and 21.8% completed primary (age 6–11 years). About 17% had no education.

Regarding occupation (Table 4.1), 40.3% of participants were in petty trading (business), 27.4% were in agriculture (farming/fishing), 10% were engineers, 5% were teachers, 5% worked as an employee of the house (domestic worker), and 2% were nurses. The majority of participants' average monthly household income lies between XAF 30,001–60,000 (USD$51.87–USD$103.74) (40.2%) and XAF 0–30,000 (USD$0–USD$ 51.87) (40%). The majority of households had about five to six people living in the home including the participant (33.2%), followed by three to four people (25.2%), and seven to eight people (17.5%). About 12% had more than 9 people living in the home. About 87% responded that there were children under five years of age living in the household.

Table 4.1 Sociodemographic Characteristics of the Respondents

Characteristics	*Variables*	*Count*	*% of Total*
Age in years	18–24 years old	28	2.83
	25–34 years old	306	30.97
	35–44 years old	358	36.34
	45–54 years old	270	27.33
	55–64 years old	25	2.53
	65–74 years old	0	0.00
	75 years or older	0	0.00
Gender	Male	416	42.11
	Female	572	57.89
Ethnicity	Arabes-Choa/Peulh/ Haoussa/ Kanuri	0	0.00
	Biu-Mandara	39	3.95
	Adamaoua-Oubangui	15	1.52
	Bantoïde Sud-Ouest	145	14.68
	Grassfields	481	48.68
	Bamilike/Bamoun	185	18.72
	Côtier/Ngoe/Oroko	39	3.95
	Beti/Bassa/Mbam	52	5.26
	Kako/Meka/Pygmé	32	3.24
	(Foreign/Other)	0	0.00
Religion	Catholic	425	43.02
	Protestant	335	33.91
	Muslim	0	0.00
	Animist	0	0.00
	Other	37	3.74
	None	191	19.33
Highest level of education	None	166	16.80
	Primary	216	21.86
	Secondary 1st cycle	293	29.66
	Secondary 2nd cycle	175	17.71
	Superior	138	13.97
Occupation	Agriculture (Farming/Fishing)	271	27.43
	Petty Trading (Business)	399	40.38
	Teaching	48	4.86
	Nursing	20	2.02
	Engineering	99	10.02
	Employee of house	50	5.06
	Other	101	10.22

Characteristics	*Variables*	*Count*	*% of Total*
Average monthly household income (in XAF)	0–30,000	396	40.08
	30,001–60,000	398	40.28
	60,001–90,000	96	9.72
	90,001–120,000	52	5.26
	120,001–150,000	46	4.66
Head of household	No	614	62.15
	Yes	374	37.85
Relationship with head of household	Spouse	599	60.63
	Brother/sister	36	3.64
	Uncle	57	5.77
	Aunt	56	5.67
	Grandparent	18	1.82
	Other (Father)	27	2.73
	Other (Mother)	1	0.10
Number of people in home, including respondent	1–2 people	118	11.94
	3–4 people	249	25.20
	5–6 people	328	33.20
	7–8 people	173	17.51
	9–10 people	59	5.97
	More than 10 people	61	6.17

Note. N = 988

Access to safe water

About 83% of respondents stated that potable water is available to them, but 50.3% stated it was not available year-round—sometimes unavailable for more than one month at a time. When asked about the reason for the unavailability of potable water, the respondents reported it to be due to damage or drying of water sources, unavailability of maintenance parts, and sometimes due to the lack of skilled workforce to fix the source. Respondents retrieved drinking water for the household from rivers (34.5%), streams (25%), springs (21.9%), and rainwater collection (21.9%). Only 16% retrieved drinking water from a public tap or standpipe. For cooking, handwashing, and other household purposes, water was most commonly retrieved from rivers (97.6%), rainwater collection (91.9%), streams (86.9%), public tap/standpipe (78.7%), and springs (78.5%). All respondents stated that they travel to their water source, with the majority stating that they walk to the source (97.1%) while some also cycled to the source (5%). About 17% of respondents stated the roundtrip from home to the water source and back takes less than one hour, while 11.8% stated it takes more than one hour. The majority of the respondents (71.26%) did not know the time taken to travel to the water source and back.

About 36% do not treat their water to make it safe to drink. Of those who do treat their water, using a water filter (ceramic, sand, composite, etc.) was the most common way to treat the water (11.3%), followed by boiling water (8.2%), letting the water stand and settle (4.3%), and adding bleach/chlorine (3.8%).

Sanitation and waste management

Of the 988 respondents, 67% of respondents have a pit latrine, 17.9% have a flushing toilet, and only about 14.5% have no facilities which necessitates traveling a distance to openly defecate near the beach or bushes. Yet, 71.7% stated that they openly defecate. About 34% do not empty their latrines or septic tanks, 63.2% stated that their septic tanks are manually emptied when they fill up, and 91.4% stated they share their toilet facility with other households.

Respondents stated that solid waste (garbage) was openly dumped (97.1%) or burned (20.2%). When asked about beliefs regarding sanitation (Table 4.2), 81% of respondents strongly agreed that having a mobile phone is more important than improving a latrine. In addition, the majority

Table 4.2 Beliefs regarding open defecation and latrines within the community

Characteristics	*Variables*	*Count*	*% of Total*
Open defecation is bad	Strongly Disagree	0	0.00
	Disagree	0	0.00
	Neutral	0	0.00
	Agree	135	13.66
	Strongly Agree	853	86.34
People who defecate in the open put their children at risk of disease	Strongly Disagree	0	0.00
	Disagree	0	0.00
	Neutral	0	0.00
	Agree	149	15.08
	Strongly Agree	839	84.92
People who defecate in the open put the whole community at risk of disease	Strongly Disagree	0	0.00
	Disagree	0	0.00
	Neutral	0	0.00
	Agree	163	16.50
	Strongly Agree	825	83.50
Open defecation is common	Strongly Disagree	0	0.00
	Disagree	0	0.00
	Neutral	0	0.00
	Agree	130	13.16
	Strongly Agree	858	86.84

Characteristics	*Variables*	*Count*	*% of Total*
In this community, a mobile phone is more important than improving a latrine	Strongly Disagree	0	0.00
	Disagree	63	6.38
	Neutral	47	4.76
	Agree	77	7.79
	Strongly Agree	801	81.07
In this community, school fees are more important than improving a latrine	Strongly Disagree	0	0.00
	Disagree	38	3.85
	Neutral	42	4.25
	Agree	77	7.79
	Strongly Agree	831	84.11
In this community, a bicycle is more important than improving a latrine	Strongly Disagree	0	0.00
	Disagree	107	10.83
	Neutral	14	1.42
	Agree	77	7.79
	Strongly Agree	790	79.96
Do you think latrines are physically safe to use?	Never	0	0.00
	Rarely	529	53.54
	Some of the time	14	1.42
	Most of the time	167	16.90
	Always	278	28.14

Note. N=988

of respondents strongly agreed that paying for school fees (84.1%) and owning a bicycle (79.9%) were more important than improving a latrine. Interestingly, 53.5% of respondents stated that they rarely felt latrines were physically safe to use.

Health and hygiene

When asked about diseases the participants may have suffered from in the past two weeks, 95.2% of respondents suffered from malaria, 74.3% suffered from diarrhoea, and 23% suffered from typhoid. When asked the same question about children under five in the household, 85.6% of children suffered from malaria, 12% suffered from diarrhoea, and 10.8% suffered from anaemia.

Respondents stated that common methods for disposing children's stool include leaving it in the open (83.9%), throwing it in the garbage (81.4%), rinsing it into toilet or latrine (79.5%), or dumping it into the ocean (26.5%). When children get diarrhoea, common methods of disposal include throwing it in the garbage (81.8%), dumping it into the ocean (77.3%), and rinsing it into the toilet/latrine (71.4%). When asked about beliefs, 86.3% of respondents strongly agreed that open defecation provides no privacy and is bad. In addition, the majority of respondents strongly agreed that people who defecate in the open put their children at risk of disease (84.9%) and put the whole community at risk of disease (83.5%). Nevertheless, 100% of respondents agreed that open defecation is common within the community.

In regard to handwashing, although 97.1% of respondents stated they do not have access to a handwashing facility, more than 90% of respondents stated they and their children wash their hands before and after eating and after using the toilet.

5 Qualitative data analysis results

The Idenau municipality conducted a community discussion of 42 participants and two one-on-one interviews with residents of the Idenau population. The community discussion was conducted in a church setting, where most participants were observing the "beginning of the year" fast. Two interviews were conducted with leaders of the Idenau municipality groups.

Throughout the qualitative responses, the three main themes that came up were lack of clean and adequate water, insufficient management of waste, and lack of hygienic behaviours in the community.

Access to safe water

Water seemed to be the most common issue brought up throughout the interviews and community discussion. Participants expressed their concern regarding the lack of available clean water. Throughout the qualitative responses, participants described their knowledge of how poor water can lead to diarrhoea and other illnesses, which can be prevented by improving access to drinking water. There is also a lack of water to properly wash their food. One participant from the community described how water storage was an issue:

> *"We store water in the house, in a very clean container, but we can store water in the house for about 2 months because . . . water in Idenau, it cannot carry the population of Idenau."*

The participant described how they continue to drink the water kept in the container for more than two months. Another participant described using containers with no lids:

> *"At times, I use containers that don't have a lid, just because I need to store water. At times I don't have a cover for the water, I just need to store the water in the house. There will be times where at three weeks I*

> *will still be using the water because there is no water. So I will use the container that doesn't have a lid."*

A third participant described that they used drums to store water, even if there were no lids, due to the lack of access to water.

Waste management

Another recurring issue was the management of waste within Idenau. Lack of education in waste management was brought up during interviews and community discussion. Individuals in the community seemed to dispose of waste carelessly, throwing it away both in the streets and around their homes. One participant described it as:

> *"It can also help to provide trash cans in the streets because the habit of disposing sewage carelessly. People just throw their sewage or garbage anywhere. So when they do this, they provide a breeding ground for insects or flies, which when they land on that sewage or garbage, they go back to houses and go on your food that you keep exposed. Those same flies come and land there."*

During the community discussion, a need for trash cans particularly in the streets in the community was emphasized. In addition, when asked if the participants would be interested in attending a training, they agreed and expressed an interest in learning how to keep their environment clean and dispose of waste properly. See Figure 5.1.

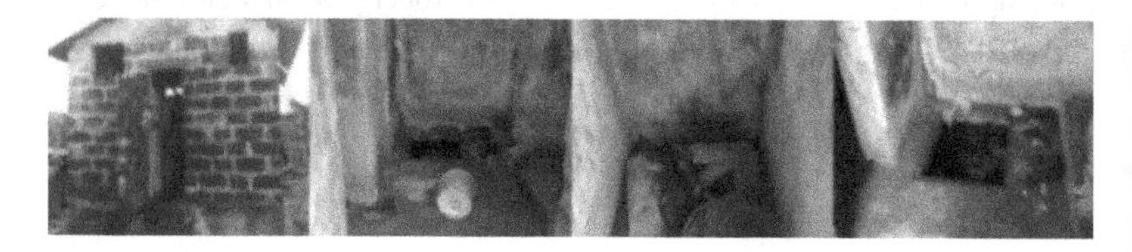

Figure 5.1 From left to right: (1) image shows the exterior of an existing communal toilet in Idenau municipality, (2) image shows four toilet stalls across each other without doors and an in-ground water tank, (3) image shows an example of a typical toilet stall filled with solid waste, and (4) image shows a closeup of the in-ground water tank from which community members are expected to scoop out water and wash after using the toilet. Without any ways to cap or separate the wash tank from the toilet stalls, hygiene is a major concern. Notably, they lack privacy, running water, and are unclean.

Sanitation and hygiene behaviours

Lack of proper hygiene behaviours and toilets in the community was brought up by individuals interviewed and participants in the community discussion. Participants described how the community lacks toilets, hence people within the community defecate in a bush near the community or in the sea. One participant from the interview describes:

> *"Ninety percent of the people in this locality defecate along the beaches and in the open roads where people move. . . . Yes, I defecate in the beach in the open because I don't have a toilet in my residence. For the cost of putting up a toilet is expensive."*

In regard to the lack of public toilets, the second interview describes:

> *"This community has a lot of challenges when it comes to this issue of public toilets. This municipality lacks public toilets. So most homes don't also have toilets, so people make use of the beaches, they make use of the streams, and they make use of the bushes. . . . As I said earlier, this issue of public toilets is a challenge."*

During the community discussion, one participant described how she has used a nylon bag or a container to defecate in and then would throw the waste into the bushes at night when nobody was looking. All participants from the qualitative responses emphasized this was an issue and understood how it affected their health by spreading diseases.

6 Discussion

Idenau, Cameroon, is a municipality on the west side of the country off the Gulf of Guinea. With a diverse population, the community is mostly engaged in subsistence agriculture and fishing. However, the community lacks sanitation infrastructure causing individuals to defecate openly throughout the community. The needs assessment aimed to understand the demographics, beliefs, WASH infrastructure, and WASH behaviours within Idenau, Cameroon, to determine the needs of the community.

Based on both quantitative and qualitative results, lack of access to clean water, open defecation and hygiene behaviours, and waste management were the main issues discussed throughout the needs assessment. Lack of clean water throughout the community forces residents to hold water in or adjacent to their homes for long periods of time in containers with or without lids, and use water from boreholes, wells, springs, and streams contaminated with sludge from oil mills and human waste. Lack of sanitation infrastructure limits proper hygiene; many participants identified the lack of public toilets, so their common method of managing excrement is by openly defecating and then disposing it into rivers, streams, bushes, or beaches. The participants agree that open defecation is bad but still reported it as a common practice within the community. In addition, malaria, diarrhoea, and typhoid are all common diseases that occur within the community. Participants in the study suggested that these diseases would be preventable with adequate water and sanitation infrastructure.

A limitation of the survey is that the responses are self-reported. While the sample size is large, the survey was implemented by volunteers from the community, which may lead to some bias in responses. Most of the questions asked in the survey were used in previous surveys in populations similar to Idenau. However, the open-ended questions asked during the community discussion and the quantitative survey questions compiled together were not assessed for validity or reliability to determine if it was appropriate for the population. In addition, conducting the community discussion proved to be a challenging task because it was during the period of maize planting, so most community members did not have time to participate in the discussion, which led to a lack of responses from the maize planting population. Some bias during the community discussion may have existed, due to the phrasing of questions by the volunteers. In addition, the large group setting that held the community discussion may have led to participants holding back their thoughts and opinions. Nevertheless, these conversations still give us insight into the current problems faced in the municipality and support the report created by the Council in 2011. Their responses bring to light the dangerous WASH situation the municipality is in.

A similar survey in 2008 was conducted by Ndjama, Beyala, Nkamdjou, Ekodeck, and Awah (2008) to understand the water supply, sanitation, and health risks in Douala, Cameroon. The study used open- and close-ended questions through interviews to review water supply sites and difficulties, dumping sites of household refuse and human waste, wastewater drainage, presence of mosquitoes, and health risks within the household. Results showed refuse was disposed of in public refuse vats, pits, in open spaces or farms. Wastewater was disposed of around houses and yards, and some in drains. Participants suffered from various diseases including cholera, diarrhoea, dysenteries, typhoid fever, malaria, and skin diseases. CAMWATER, the National Company of Water of Cameroon that is the mass distribution firm for supplying reliable drinking water in Cameroon, provides clean water and has increased access to water in communities in Douala. However, the level of access to CAMWATER drinking water increased with an individual's income. There was a strong use of public refuse vats, but poor management of human excrement still existed due to the lack of knowledge of the consequences of these behaviours within these communities. Interestingly, our study contradicts this and shows individuals in the Idenau municipality identified that open defecation was bad and puts individuals at risk; however, it was still common in Idenau. Health risks found in this study were similar to that found in our needs assessment, emphasizing that diarrhoea, malaria, and typhoid were common preventable diseases found in the Idenau municipality.

Bibliography

Ndjama, J. P., Beyala, V. R. K. K., Nkamdjou, L. S., Ekodeck, G., & Awah, T. M. (2008). Water supply, sanitation and health risks in Douala, Cameroon. *African Journal of Environmental Science and Technology*, *2*(12), 422–429. Retrieved from www.academicjournals.org/AJEST

7 The UN Sustainable Development Goals (SDGs)

The United Nations 2030 agenda for sustainable development

In furtherance of its mission to eradicate poverty in all forms around the world, the United Nations (UN) created 17 Sustainable Development Goals (SDGs). These goals were established in order to promote human rights and gender-based equality globally by 2030. The goals focus on equality for all people, protecting the planet, economic prosperity, and fostering peaceful, inclusive societies. Countries around the world have committed to implementing the agenda within their own borders and to reach these goals through international cooperation. Each country faces unique challenges based on its respective geographic and demographic situation.

The inclusion of the Sustainable Development Goals is vital in planning and developing the Idenau municipality. The framework allows us to ensure all interventions and recommended actions for the future development of the municipality are inclusive and take into consideration all individuals in the municipality. The challenges and needs addressed in this book are not unique to Idenau – indeed these challenges are common to many communities around the world.

For the context of the Idenau municipality, several SDGs are particularly relevant. **SDG 3** is to "Ensure healthy lives and promote well-being for all ages." SDG 3 comprises nine targets that focus on reducing mortality rates for vulnerable populations, reducing risk factors for communicable and non-communicable diseases, providing universal health coverage, and as a result, strengthening the health sector (UN, 2015).

Although significant progress has been made over the last decade to curb the disease burden, malaria still remains the leading cause of mortality and morbidity in Cameroon among vulnerable groups including children under five and pregnant women (Antonio-Nkondjio et al., 2019). In Idenau, communities are suffering from health problems such as malaria, diarrhoea, and typhoid due to a number of reasons including, but not limited to, lack of access to clean water, lack of proper sanitation, and inadequate housing. People of all ages, including children under five, are struggling to survive from health concerns that are easily preventable. Malaria, diarrhoeal diseases, and lower respiratory infections are part of the top five most common, yet preventable, causes of death in Cameroon and among the top ten causes of death in low-income countries (WHO, 2018). Atieli, Menya, Githeko, and Scott (2009); Cattaneo, Galiani, Gertler, Martinez, and Titiunik (2009); Lwetoijera et al. (2013); Yé et al. (2006) show that interventions at the intersection of health and adequate housing create significant reduction in preventable infectious diseases through simple modifications to the existing housing structures such as screening doors, windows and eaves to prevent malaria and replacing dirt floors with a cleanable flooring system to prevent diarrhoea-related diseases.

Every country has a unique national poverty line. The latest data available from 2014 show that the population living below the national poverty line in Cameroon decreased to 37.5% of the population from 40% in 2001, with 24% of the population living on $1.90/day from 2001 to 2014 ("Poverty headcount ratio at national poverty lines (% of population) – Cameroon | Data," 2020). A majority of the current population in Idenau is currently living on $1.90/ day or less. This poses a growing risk to their access to water and sanitation services because social determinants of health such as their economic status are directly linked to their ability to access these facilities and vice versa.

SDG 6 relates to the importance of water and sanitation. The framework strives to "ensure availability and sustainable management of water and sanitation for all" with five targets to increase access to safe drinking water, end open defecation, and increase access to sanitation facilities including handwashing stations, and waste management. Water, sanitation, and hygiene are essential for good health and well-being. All individuals around the world should have equitable access to clean, safe, affordable drinking water. According to the WHO/UNICEF Joint Monitoring Programme for Water Supply, Sanitation and Hygiene (JMP), a basic drinking water facility is defined as having access to drinking water from an improved source, provided the roundtrip collection time, including queuing, is 30 minutes or less. The latest data available from 2017 showed 60.4% of people in Cameroon had access to basic drinking water services, and 9.4% have basic handwashing facilities including soap and water. Statistics from 2015 showed that 39% of the population had access to basic sanitation services, which has gradually increased since then. However, more and more people have been migrating to urban slums in search for better economic opportunities where their ability to access these facilities is less reliable. Improving the sanitation infrastructure can not only be a critical component to ending open defecation among residents, but also in providing clean, drinkable water for those in Idenau. Many of the SDG goals and targets are interlinked to each other. SDG 6 is directly related to SDG 3, and an increase in access to clean water can prevent diarrhoeal diseases and other illnesses from developing. Despite the gradual decrease in the under-five child mortality rate in Cameroon from 116.30 per 1,000 live births in 2008 to 76.10 per 1,000 live births in 2018, areas like the Idenau municipality remain unnoticed and continue to suffer from disease-related morbidity and mortality. Sanitation-related activities and programmes, such as treating wastewater, recycling, and waste disposal are also important aspects of SDG 6 that have a significant impact on health. A lack of sanitation infrastructure can act as a barrier to health and longevity.

SDG 11 focuses on the built environment. The framework was developed to "make cities and human settlements inclusive, safe, resilient and sustainable." Making the built environment safer and healthier for people can not only help lower-income individuals lead prosperous lives, but can also enable future generations to succeed and flourish. For the overall health of the community through the built environment, homes must be designed to support health. Making upgrades to individual homes will have a ripple effect making the community safer and more resilient in the long run. ARCHIVE's work focuses on improving housing in vulnerable communities to support the health of their residents and has a proven track-record of strengthening its surrounding community. Inadequate housing includes open eaves, unscreened windows and doors, and a lack of a ceiling barrier with roofing material. These are all factors that allow mosquitoes to easily enter the indoor living environment and transmit malaria. Mud floors have the capacity to harbour bacteria and parasites and are nearly impossible to clean, putting children in the household, especially toddlers, at risk of diarrhoeal disease, respiratory and skin infections. Simple modifications to existing homes, such as sealing eaves, directly affects the number of mosquitoes entering the home, thereby decreasing the risk of infection by malaria-carrying mosquitoes for all residents. ARCHIVE has worked on several screening projects for malaria reduction in communities across Africa. An example of this is the project Building Malaria Prevention, in Yaoundé, Cameroon, which catalyzed a 20% decrease in incidences of malaria and a 50% reduction in exposure to malaria-carrying mosquitoes in participating homes. ARCHIVE's Mud to Mortar project in Bangladesh replaces dirt floors with durable, cleanable, and cost-effective concrete floors to combat diarrhoea-related diseases. In addition to implementation, ARCHIVE also collects information about the family's health before and after the intervention as part of a research study and disseminates an education campaign so that together these components can provide evidence of the impact of the built environment on health conditions.

SDG 13 emphasizes the need for countries to "take urgent action to combat climate change and its impacts*." The asterisk refers to the "the United Nations Framework Convention on Climate Change (UNFCCC) as the primary international, intergovernmental forum for negotiating the global response to climate change." The consequences of climate change affecting Cameroon include abnormal recurrences of extreme weather changes such as violent winds, high temperatures, and heavy rainfall, which cause widespread damage to the built environment. Flooding remains a considerable problem in the Idenau municipality due to deforestation, poor drainage systems, and changes in weather patterns. Due to poor knowledge of forest rules and regulations, there is a vast industry of illegal exploitation of timber and deforestation for farming activities. These practices cause seasonal changes such as an increase in temperatures, change in precipitation patterns, change in extreme weather events, and reduction in the availability of water, which in turn affect the agricultural industry. Cameroon's infrastructure struggles to keep up with these climatic changes: increased flooding from heavy precipitation leads to the destruction of property, increased stagnant water in open pipes, trash, and debris increases mosquito breeding and the spread of malaria. Vulnerable populations, such as those of specific age groups like children and the elderly, low socioeconomic status, and people with certain health conditions, suffer disproportionately from climate change, especially when exacerbated by substandard living conditions.

The built environment stands at the intersection of climate change and health. Simple design interventions can help create resilient communities that better withstand crises. Targeted efforts to modify existing conditions can lower the reconstruction costs of buildings while maintaining the social fabric of a community and reducing the environmental impact of new, large construction projects after a crisis has occurred.

Understanding the implications that climate change, health, and the built environment have on each other can empower authorities to better prepare communities for change and crises. Multi-sectoral partnerships are instrumental in improving the health outcomes of vulnerable communities worldwide. This allows for a holistic approach, which is necessary in planning strategies of development for places like Idenau while taking into account the different complexities and specificities of a project and including professional experts from different sectors. **SDG 17** focuses on partnerships for the goals. One of the objectives for the Idenau municipality Development plan in 2011 was to collaborate with neighbouring councils, institutions, stakeholders, and forest and wildlife conservation partners (*Idenau Council Development Plan*, 2011). ARCHIVE aims to support the Idenau municipality by introducing and forming multi-sectoral and multi-regional partnerships that are critical for achieving the Sustainable Development Goals for Cameroon.

Cameroon – the 2019 SDG Index score and rank

Since 2015, the 2030 Agenda has been a map to a sustainable planet. Each year the United Nations publishes the Sustainable Development Report providing a detailed account of each country's progress and shortcomings, including the global progress made and challenges that remain to achieve the goals by 2030. While significant progress is being made, the evidence shows that there is a dire need for urgent attention to accomplish the SDG goals for the 2030 Agenda. The Sustainable Development Report presents the SDG Index and Dashboards to summarize a country's current performance, strengths, and weaknesses in their progress. The SDG score shows a country's position on the outcomes between the worst (with a score of 0) to the best (with a score of 100). In the 2019 SDG Global ranking, Denmark ranked number one with the highest index score of 85.2, which suggests the country is at 85.2% of achieving all SDG targets. From the SDG Report of 2019, Cameroon ranked 127 of 162 countries with a global index score of 56, implying that Cameroon is about 56% of the way to achieving the SDGs – 4.1% higher than the regional average score of 53.8 for Central Africa (Sachs, Schmidt-Traub, Kroll, Lafortune, & Fuller, 2019).

According to Sachs et al. (2019), Cameroon's assessment for **SDG 3** (Good Health and Well-being score of 42.7), **SDG 6** (Clean Water and Sanitation score of 52.5), **SDG 11** (Sustainable Cities and Communities score of 29.8), and **SDG 17** (Partnerships for the Goals score of 34.2) shows that the country's scores for the goals are stagnant and lagging behind considerably. There are major challenges that remain to be accomplished, as outlined previously. With few resources, Idenau is particularly behind. Although Cameroon is on track to achieve the 2030 goal for **SDG 13** (Climate Action score of 97.9) with lower emissions of carbon dioxide per capita from the consumption of energy such as petroleum, natural gas, coal, and also from natural gas flaring, significant challenges remain. Much still needs to be done to reduce the yearly average rates of deaths, injuries, homelessness, and need for basic necessities for survival due to climate-related disasters (436.4 per 100,000 people, 2018). Similarly, the residents of Idenau continue to be impacted by climate change. There is a need for direct action to educate the community about ways to mitigate and prevent some of the effects they are experiencing.

The agenda for 2030 Sustainable Development encourages member countries to submit a voluntary report on the SDGs with the aim of facilitating dialogue on experiences of each country in achieving the goals, including their successes, challenges, and lessons learned. The government of

Cameroon submitted its first Voluntary National Review (VNR) in 2019 on the country's efforts to reach the Sustainable Development Goals. The report speaks to SDGs 4, 8, 10, and 13. According to Cameroon's VNR, the government of Cameroon is currently working on a new strategy for the education sector that aligns with the SDG 4's Quality Education targets. The report also states that in order to achieve the goal of standard education for all, the challenge faced by Cameroonians is inadequate school infrastructure and an insufficient number of qualified teachers in the country. In the fight against climate change, Cameroon has agreed to the Paris Agreement and to prepare a national REDD+ strategy to reduce emissions from deforestation and forest degradation to reduce the Greenhouse Gas emissions by 32%. Cameroon has also attempted to address climate change through the Agricultural Investment plan; however, they face a major challenge in financing the alternative activities to eliminate deforestation, which is by far the main factor for greenhouse gas emissions.

The 2019 VNR report suggests that Cameroon plans to focus on the country's economic growth, the education sector, reducing inequalities, climate change, policies, and mobilizing resources and partnerships. ARCHIVE aims to align with this plan on several points such as reducing inequalities, reducing the effects of climate change, and mobilizing resources and partnerships. More specifically, the goal for the Idenau municipality is to use the SDGs to improve health outcomes by recommending and implementing targeted interventions to address SDGs 3, 6, 11, 13, and 17.

The SDGs set benchmarks that city officials, community-based organizations, and NGOs can use as a reference in creating a comprehensive WASH plan for the municipality. SDGs set the standard and the objectives towards inclusive development, and an increased focus on these can be leveraged to gain partnerships and aid support from the international donor community. With the help of these partnerships, the project can attain safer and healthier living standards for the people of Idenau. The 2019 SDG report highlights the lack of investment in data collection by most countries on more than half of the global indicators. The lack of accurate data on marginalized populations makes them increasingly invisible and exacerbates their vulnerability. Therefore, collecting data to accurately represent the conditions within Idenau before and after the interventions will be critical in helping to show how small but growing urban centres can improve WASH conditions while aligning with the SDGs.

Government action, 2010–2020

The 2008 international financial, food, and energy crisis reached Cameroon and resulted in a rise in the cost of living, causing multiple riots within the country. To take action, the authorities revised the Poverty Reduction Strategy

Paper published in 2003 and renamed it the Growth and Employment Strategy Paper, involving people at the grassroots level and using a participatory planning approach. The authorities published the Growth and Employment Strategy Paper (GESP) in 2009 as a reference framework for government action over the period of 2010–2020. The GESP reviewed the development policies of Cameroon's economic and socio-economic situation, challenges faced by major sectors, opportunities and threats and visions and goals to be reached by 2035. These include a growth strategy for infrastructure development, production mechanism, human development, trade, economy; employment strategies; state governance and strategic management; macroeconomic and budgetary framework; and a framework to monitor the implementation of the GESP.

The 2035 vision and goals consist of the following general objectives:

1 Reduce poverty to a socially acceptable level
2 Become a medium-income country
3 Become a newly industrialized country
4 Strengthen national unity and consolidate democracy by promoting the ideals of peace, freedom, justice, social progress, and national solidarity

As a part of these objectives, the government of Cameroon committed to several infrastructural developments including maintenance and rehabilitation: construction of road networks, telecommunications infrastructure, urban housing, and an increase in the production of electricity enabling Cameroon to export electricity, thus boosting the country's economic growth. The government also plans to reduce morbidity by a third among the poor and mortality by two-thirds among children under the age of five. They plan to do so by increasing the rate of access to safe drinking water by 75% in 2020 by providing 40,000 new water points, renovating 6,000 existing water points in rural areas, as well as 1,200,000 latrines, thereby increasing the population's access to basic sanitation needs (IMF, 2009).

Cameroon's progress in the implementation of the SDGs

The United Nations introduced an implementation survey aimed to understand how each member country's current governments are implementing the Sustainable Development Goals. The survey covers six strategic aspects: (1) national strategies and baseline assessments, (2) budgeting practices and procedures, (3) stakeholder engagement, (4) coordinating units, (5) legislative actions, and (6) main challenges for implementation. The results show the government of Cameroon has made some attempts at implementing the

SDGs. For example, an official statement was made by a high-ranking official endorsing the SDGs and the country's plan to put the implementation of SDGs in effect. However, there is no evidence that the SDGs are being integrated into sectoral action plans. Cameroon has clearly defined the key national priorities and has included 217 of the 231 unique SDG indicators to their list of official national indicators to be achieved. However, they have not issued any report about SDG progress and implementation of strategies to meet them, apart from their first VNR report in 2019. The VNR reported that an assessment was made on how far Cameroon is from achieving all the SDGs but failed to provide a quantitative measure of the distance remaining to reach the targets. The implementation survey also shows that Cameroon has not assessed the finances needed to achieve the SDGs and shows no plans to do so, as there is no mention of the SDGs in their latest national budget. The Cameroon government has created a central unit under the Ministry of Planning that is responsible for supporting the implementation of SDGs across different departments and agencies, but has yet to take legislative actions to operationalize the implementation.

The government of Cameroon has made attempts to address the challenges faced by its population by setting goals to achieve the vision of emerging as a middle-income country by 2035. Addressing limitations and setting goals are important for change, but an implementation strategy is needed in order to achieve the vision of a middle-income country by 2035 along with the 2030 Agenda for Sustainable Development for all.

Bibliography

Antonio-Nkondjio, C., Ndo, C., Njiokou, F., Bigoga, J. D., Awono-Ambene, P., Etang, J., . . . Wondji, C. S. (2019). Review of malaria situation in Cameroon: Technical viewpoint on challenges and prospects for disease elimination. *Parasites & Vectors*, *12*(1). https://doi.org/10.1186/s13071-019-3753-8

Atieli, H., Menya, D., Githeko, A., & Scott, T. (2009). House design modifications reduce indoor resting malaria vector densities in rice irrigation scheme area in western Kenya. *Malaria Journal*, *8*(1), 108. https://doi.org/10.1186/1475-2875-8-108

Cattaneo, M. D., Galiani, S., Gertler, P. J., Martinez, S., & Titiunik, R. (2009). Housing, health, and happiness. *American Economic Journal: Economic Policy*, *1*(1), 75–105. https://doi.org/10.1257/pol.1.1.75

Idenau council development plan. (2011). Retrieved from PNDP, Southwest Regional Coordination Unit: www.pndp.org/documents/10_CDP_Idenau.pdf

IMF. (2009). *Cameroon : Poverty reduction strategy paper*. Retrieved from Washington, DC: www.imf.org/en/Publications/CR/Issues/2016/12/31/Cameroon-Poverty-Reduction-Strategy-Paper-24119

Lwetoijera, D. W., Kiware, S. S., Mageni, Z. D., Dongus, S., Harris, C., Devine, G. J., & Majambere, S. (2013). A need for better housing to further reduce indoor malaria transmission in areas with high bed net coverage. *Parasites & Vectors*, *6*(1), 57. https://doi.org/10.1186/1756-3305-6-57

Poverty headcount ratio at national poverty lines (% of population) – Cameroon | Data. (2020). Retrieved from https://data.worldbank.org/indicator/SI.POV.NAHC?locations=CM

Radical increase in water and sanitation investment required to meet development targets. (2017). [Press release]. Retrieved from www.who.int/en/news-room/detail/13-04-2017-radical-increase-in-water-and-sanitation-investment-required-to-meet-development-targets

Sachs, J., Schmidt-Traub, G., Kroll, C., Lafortune, G., & Fuller, G. (2019). *Sustainable development report 2019*. Retrieved from New York: https://sdgindex.org/

UN. (2015). *Transforming our world: The 2030 agenda for sustainable development*. Retrieved from https://sustainabledevelopment.un.org/content/documents/21252030%20Agenda%20for%20Sustainable%20Development%20web.pdf

World Health Organization: WHO. (2018). The top 10 causes of death. Retrieved from https://www.who.int/news-room/fact-sheets/detail/the-top-10-causes-of-death

Yé, Y., Hoshen, M., Louis, V. R., Séraphin, S., Traoré, I., & Sauerborn, R. (2006). Housing conditions and Plasmodium falciparum infection: Protective effect of iron-sheet roofed houses. *Malaria Journal*, *5*. https://doi.org/10.1186/1475-2875-5-8

8 Actionable recommendations

In response to the municipal direction and the survey's qualitative responses, three main themes around sanitation were consistent for the Idenau municipality: there is a lack of consistent access to clean and adequate potable water, a lack of knowledge of healthy hygiene behaviours, and little to no waste management. As highlighted here, a key priority will be community education to ensure that interventions made will be understood, supported, maintained, and utilized. Another priority for long-term sustainability will be the creation of an industry and employment around waste management systems.

The central government is committed to improving access to safe drinking water and basic sanitation infrastructure by 2020 through three key measures: rehabilitating existing infrastructure that has fallen into disrepair, expanding existing potable water networks where they exist, and creating extended systems that can reach remote areas. It is not clear at this moment exactly which benefits the Idenau municipality will reap from some of these measures as the intent is to improve 75% by 2020 nationally. So full coverage may benefit certain geographical regions more than others. The actionable recommendations are to be discussed in detail with the authorities of the municipality to understand if Idenau is expected to benefit from this plan and to dovetail it into the recommended actions that follow.

The actions proposed by the project can be divided into several phases that build on each other. The intent of outlining the project in this way is to ensure that certain steps are taken before others, so that there is an opportunity to learn and build on each phase. In addition, the municipality has stated that it does not have funds to dedicate towards improving WASH conditions in the community, so resources may be more accessible if divided into clearly defined work packages. The findings from the needs assessment have proved that much about this community and the challenges they face is still not fully known. The pilot phase is critical in helping ARCHIVE explore that gap and provide the opportunity to find answers to the questions that have surfaced after analyzing the data collected for the needs assessment. Hence, the suggested way forward is to implement interventions by phases to learn and improve strategies in each phase.

Pilot phase – estimated timeline 12 months

As a starting point, a comprehensive baseline survey is necessary to understand how each initiative and each round of implementation impacts the community. The Pilot Phase will include a sensitization exercise to explore the topic of open defecation with the idea that embedding knowledge and awareness around why the changes are important will be critical to long-term success. It will also generate additional demand from the community for the improvements to come in further stages. Lastly, introducing the concept of a sanitation service chain and identifying key stakeholders would begin in the Pilot Phase as learning about opportunities and potential challenges to this initiative will be critical in outlining how to fund the efforts to develop it further. The following are the actionable recommendations for the Pilot Phase that have been approved by the mayor:

1. **Comprehensive baseline data collection.** This includes:
    a. an in-depth qualitative study among community members and municipality authorities to understand waste management strategies and perception,
    b. a baseline health survey to track changes in the community after each phase, and
    c. a windshield survey to determine the location and conditions of existing infrastructure, refine the map of the community, and catalogue/survey existing housing stock.
2. **Community sensitization campaigns.** A training and education campaign on sanitation, hygiene behaviours and environmental education, including a Community-Led Total Sanitation (CLTS) campaign to address open defecation. This includes:
    a. awareness building exercises within the community,
    b. reaching a consensus that sanitation and stopping open defecation should be a community-wide priority,
    c. identifying champions of the cause to encourage long-term change, and
    d. guiding community members to identify and use locally available resources to build their own toilets in the community.
3. **Feasibility analysis to develop a sanitation service chain.** This will include using an excreta flow diagram to analyze the following:
    a. access to toilets,
    b. pathways for safe removal of faecal waste,
    c. modes of transportation of faecal waste,

   d. whether or not faecal waste is treated, and
   e. modes of disposal/reuse of by-products.

4. **Stakeholder engagement.** Engage with the municipal authority, community members, and entrepreneurs for input, and approval of implementation.

Potential barriers to implementation

1. Identifying and training enumerators – training will be required, and we may need them to travel from a different community. In light of COVID-19 and the political situation in Cameroon, this barrier may be a significant challenge.
2. Acquiring or creating an accurate, detailed map of the community to enable the design of appropriate interventions may prove difficult unless students or young professionals from Architecture or Engineering professions are identified and trained to conduct the surveys.
3. Diverse sources of funding or larger grants with holistic values are required to address the diverse actionable recommendations.
4. The CLTS effort may encounter challenges during the triggering phase, the implementation of self-constructed toilets, and achieving ODF status.

Phase I – estimated timeline 24 months

The intent of Phase I is to build on the successes of the Pilot Phase and/or adjust efforts where needed. The goal of the Pilot Phase is for the municipality and their champions to achieve the ODF status and identify ways to support this effort and evaluate the options available to develop a sanitation service chain. In Phase I the intent is to use community participatory planning exercises to identify and implement necessary measures to facilitate maintaining the ODF status and to develop a waste management plan for the municipality. The following are the actionable recommendations for Phase I that have been approved by the mayor:

1. **Maintaining and expanding on CLTS efforts.** To facilitate maintaining ODF status within the community, participatory planning exercises with the members of the community and the municipal authorities will be conducted to identify the necessary support for maintaining and improving the facilities created by the community in the CLTS effort. The success of these interventions is key in implementing the sanitation service chain. Depending on the community participatory planning exercise:

   a. additional communal toilets and
   b. handwashing stations may be added.

2. **Initial waste management plan.** A waste management intervention based on the findings from the Pilot Phase and the community participatory planning exercises will be proposed. This may include:

   a. trash cans along the streets and in key locations,
   b. a trash collection system,
   c. advocacy campaign around waste management, and
   d. understanding of how some of these components will fit into the sanitation service chain effort.

3. **Phase I impact evaluation.** A post implementation evaluation of Phase I impacts on health improvements, behaviour changes, and knowledge retention will be conducted to compare changes from the baseline.
4. **Stakeholder engagement.** The project will continue to engage with the municipal authority and community members for input, community engagement, and approval of implementation at regular intervals.

Potential barriers to implementation

1. The implementation of Phase II, especially the success of the sanitation service chain, is dependent on the municipalities' ability to maintain their ODF status.
2. Identifying an entrepreneur and adequate loans to launch the sanitation service chain.
3. Identifying diverse sources of funding, or sources that are interested in holistic project approaches to address the varied actionable recommendations.

Phase II – estimated timeline 36+ months

The intent of Phase II is to put the separate pieces together to create a holistic initiative to combat the lack of access to potable water, to find solutions around waste and sanitation, and to create a healthier community from the household level to the city scale. A self-sustainable business that handles waste could support the municipality to advocate for industrial waste to be disposed of more appropriately. This in turn will allow the inhabitants to be healthier; have access to potable water, which will enable parents to be able to go to work more days of the year and children to attend school more regularly. The following are the actionable recommendations for Phase II that have been approved by the mayor:

1. **Increase access to adequate housing.** The interventions to increase access to adequate housing will be identified through a community

participatory planning effort. Efforts will be made to identify the most pressing needs of the community ensuring the families are well equipped to combat diseases, gain access to improved health and develop communities that are more resilient. Some examples of the interventions are:

a. Upgrade dirt floors to combat diarrhoeal diseases at the individual household level as a pilot project,
b. Screen windows, eaves, and doors to reduce malaria incidence in the community, and
c. Ventilation of the homes to reduce indoor air pollution to reduce respiratory illnesses.

2. **Create access to potable water supplies.**

   a. Evaluation of existing infrastructure to see if it can be restored.
   b. Protect and capture clean water sources nearby.
   c. Regular (every six months) bacteriological and chemical water surveillance to ensure the water remains potable over time.
   d. Create a pressurized pipe system to enable potable water to reach communities to ensure regular and sufficient access.
   e. Ensure ongoing maintenance of water facilities through privately owned small businesses or as government run projects.

3. **Advocacy.** Create necessary policies to enable long term sustainable changes within the municipality.

   a. Industrial waste – introduce enforceable policies by working with companies to ensure industrial and toxic waste is disposed of properly

4. **Phase II impact evaluation.** Conduct a post-implementation evaluation of Phase II impacts on health improvements, behaviour changes, and knowledge retention to compare changes from the baseline and Phase I.

5. **Implement a waste management service chain (Economic system).** The sanitation and waste management service chain includes:

   a. maintenance of toilets,
   b. safe removal of faecal waste using manual or mechanical techniques,
   c. transportation of faecal waste, either using smaller containers that can travel into dense residential areas and/or with a truck,
   d. treatment of faecal waste at an offsite facility, and
   e. proper disposal/reuse of by-products (mulch or other).

6. **Stakeholder engagement**. Engage with the municipality's authority and community members for input and approval of implementation

Potential barriers to implementation

1. Implementing a sanitation service chain and ensuring its long-term self-sustainability without continued financial support may jeopardize the success of better sanitation in Idenau, and its open defecation free status
2. Identifying diverse sources of funding or sources that are interested in holistic project approaches to address the varied actionable recommendations

9 Opportunities and barriers in implementation

In this chapter, we outline the factors perceived as being opportunities and impediments to the implementation of the actionable recommendations previously outlined.

Opportunities in implementation

Multi-sectoral partnerships

Since its inception, ARCHIVE has conducted its work by building partnerships, working with in-country local partners, stakeholders, government representatives, designers, and many others. These partnerships support communities by enabling the implementation of interventions at the intersection of health and the built environment, combating disease through simple modifications to existing structures. ARCHIVE aims to create a multi-sectoral team in order to achieve the actionable recommendations with the Idenau municipality. The team will include experts from the private and public sectors who will assist in planning and implementation. This will also strengthen the engagement efforts, allowing the team to understand the most critical needs within the community that the project can respond to. The aim is to identify participants from key sectors, both locally and internationally, such as public health professionals, WASH and waste management infrastructure experts, environmental experts, research, monitoring and evaluation professionals, design and construction professionals, and local entrepreneurs and local businesses.

Local stakeholder engagement

The initial and most important step is to identify key local stakeholders in the early stages of planning. The engagement and consistent communication with Idenau's local representative to conduct the needs assessment was fundamental to ensure the voice of the community was heard. The stakeholders in the Idenau municipality include the authorities from the municipality, religious institutions, health clinics, social workers, youth groups, local community organizations, and leaders of community groups. These stakeholders represent different community interests, can bring local resources to support the interventions, and are committed to developing the community. The opinions and insights from these groups of people are valuable throughout the entire process of the intervention including the planning, implementation, and evaluation of the intervention. They are the key partners in mobilizing communities for the CLTS and community participatory planning exercises.

Local authority as a core partner

The local authorities are democratically responsible for the well-being of the individuals in their constituency. Since the municipality authorities in Idenau are accountable for spearheading all initiatives for the betterment of the community and are in direct contact with the residents, it is only natural that they play a critical role in the development plan of the Idenau municipality. With this motivation ARCHIVE has presented the findings of the needs assessment and the recommendations for the development of the municipality to the mayor of Idenau. The local mayor is also key in accessing resources from the region and the national government, so ensuring buy-in and support from the local authorities has been vital to moving the project forward.

Project is replicable

The assessment of the issues faced by the residents of the Idenau municipality was a learning process for ARCHIVE – from facing the challenge of conducting a literature review on a community with little accurate data to working with a municipality that has very scarce resources but high needs. There is no one-size-fits-all approach; each community has its own priorities and faces different issues. As outsiders, we need to understand the local context first before planning interventions. Being an external facilitator, over the years, ARCHIVE has successfully tailored interventions based on local, cultural, and social needs of a community facing similar challenges compared to communities in other countries. ARCHIVE believes that the process of working with, designing for, and implementing with this community will be replicable in similar communities in the future.

Dissemination of data

ARCHIVE's aim in publishing the work about the Idenau municipality is to shed light on the Idenau municipality, its needs and potential; to increase the understanding of how important the built environment is for health; and to suggest a model which may be replicated in other similar communities to improve people's quality of life. The limited publications of peer-reviewed journals on communities similar to the Idenau municipality shows a global lack of interest, lack of funding, and lack of presence of Idenau and similar communities on the global research stage. Compared to the 1,100 scientific and technical journal articles published for every one million individuals in the Organization for Economic Co-operation and Development (OECD) member countries, there were only seven scientific and technical journal articles published in 2013 for every one million individuals in the low-income countries of Africa (Utoikamanu, 2019). ARCHIVE had the opportunity to identify, assess, analyze, and document the challenges and opportunities faced by the residents of this municipality, allowing ARCHIVE to exchange the knowledge gathered from this process with other experts in the industry who are working in communities similar to the Idenau municipality.

Barriers in implementation

Lack of accurate data: a common occurrence

In 2017, ARCHIVE was commissioned to conduct a needs assessment on the Idenau municipality. In the preliminary stages while collecting secondary data from the publicly available records of the community, ARCHIVE discovered the lack of peer-reviewed journals, articles, and up-to-date, reliable information such as population size or a geographical map of the Idenau municipality. The lack of accurate health data available on the community was another impediment during secondary data collection, a common obstacle in the Global South. Hence, it is critical that the findings are shared with others in the field of Global Public Health and Environmental Health to promote the need for interventions in the Idenau municipality in Cameroon and communities like it around the world.

Lack of political and budgetary priority

The prospects of economic growth in Cameroon have remained tantalizingly close to becoming a reality in the last few decades. Driven primarily by the perception of its relative political stability and the country's abundant natural resources, financial forecasters and economists have often identified the country as being on the cusp of an economic boom. A common sentiment shared by experts is that Cameroon has many advantages that would typically promote robust growth and development. This assessment is reflected in the ambitious economic goals that the federal government has outlined for the country on an annual basis. National priorities emphasize aggressive government investment in development, particularly around creating a supportive infrastructure for industry. Thus, the building of reliable electrical power grids, nationwide road networks, and industrial zones to attract foreign direct investment has featured prominently in government planning for many years.

However, the goal of increased economic prosperity is yet to be fully realized. The all-too-common challenges of systemic corruption, poverty, and uneven distribution of resources remain stubborn barriers to creating a responsive and resilient economy for all Cameroonians. Nowhere are these issues more apparent than in the restive northern states of Cameroon, which remain less developed and poorer than other parts of the country. In this area of the country, the majority of people are English-speaking and historically disenfranchised compared to their neighbouring French-speaking states. As a result, poverty has translated into increasingly vocal protestations and a cycle of violence in some areas. These issues in turn have perpetuated a cycle of neglect and violence that has exacerbated the alienation felt by English-speaking Cameroonians and the distrust of the region by officials in the federal government.

Further complicating this situation is the challenge of allocating the limited government assistance the region receives. Competing interests to fund numerous development priorities often has meant that water and sanitation projects lose to other social sectors funding needs such as education and health. This is the hard calculus for state and local officials to make. This climate of cyclic poverty, underdevelopment, disenfranchisement, and unrest makes it difficult for those who are serving the interests of the people living in municipalities such as Idenau.

Lack of resources, management, and coordination

In addition to the lack of access to necessities and challenges faced by low-income communities like in the Idenau municipality, another limitation for these communities is the gap in technology. Technological resources play an important role in bringing about change and development and are readily available globally in high-income communities unlike the Idenau municipality. An example of the challenges ARCHIVE faced during the data collection for the needs assessment was having to make multiple attempts to access a map of the municipality due to a lack of digitized maps on platforms like Google Maps and Google Earth, and missing resources enabling the digitization of the physical maps possessed by the municipality. ARCHIVE was able to attain low-quality collaged photographs of two maps of Idenau that were partially labelled by hand. The holistic nature of the actionable recommendations for the development of Idenau makes it a challenge to seek funding since most funding opportunities focus on specific priority areas.

Funding strategies

The Idenau municipality receives revenue from operations such as tax revenue, additional council tax, local developmental tax, proceeds from council taxes, proceeds of the management of land and services, rebates and royalties granted by the state, financial income, and operating allocations and investments, such as allotment fund, reserved fund, capital and investment grants, and investment refund. Due to limited information on revenue budgets, it is difficult to strategize an approach with the authorities in the municipality and federal government.

The hope is that the municipality will allocate annual funding towards portions of the strategies outlined previously and that the federal government will have improved access to clean water by 2020 as stated. Unfortunately, there are no signs of this being implemented in Idenau by the time this work is published, so it is likely that there will be a need for additional fundraising.

Given the lack of regular and substantive federal investment, municipal officials and community organizations often feel that they do not have options to turn to for technical and financial support. In an effort to propose a pragmatic solution, ARCHIVE suggests that, based on the present situation, holding out for federal government assistance may not be feasible in the short term. Instead, a multi-stakeholder coalition could be a model that may yield better results. This coalition would consist of local community-based organizations, youth groups, municipal officials, and international NGOs and foundations. Together, they could identify and accomplish achievable objectives (such as the implementation of neighbourhood-based Community Led Total Sanitation initiatives) and gradually build towards more ambitious sanitation infrastructure objectives. This multi-stakeholder coalition approach benefits from community-based momentum and outside expertise and funding. Accomplishing this is no easy task. It will take time for citizens from Idenau to champion and build networks of supporters at all levels of the community, and global advocates to persistently champion the cause of Idenau to their networks.

Bibliography

Utoikamanu, F. (2019). Closing the technology gap in least developed countries. *UN Chronicle*, *55*(4), 35–38. https://doi.org/10.18356/3a542c74-en

Conclusion

Communities with similar development challenges to Idenau's – rapidly urbanizing towns with a large proportion of self-built housing, inadequate provisions of infrastructure, lack of health literacy, and increasingly affected by climate change – can be found worldwide. The lack of accurate data on marginalized populations makes them increasingly invisible and exacerbates their vulnerability. This publication aims to demonstrate that engaging with local representatives and organizations to gather information to create meaningful change in their communities is possible. The needs assessment conducted has served to highlight the issues that require immediate attention to allow this community to thrive. Some of these issues identified require coordination across stakeholders and may only require modest resources to begin the cycle of improving the community. The information gathered so far should serve as the springboard for partners to come together with the community to define priorities, opportunities, challenges, and most importantly identify the existing resources that they can leverage for lasting change. The question of resources is well illustrated with the example of water in the municipality, there is sufficient water from natural sources, but it is being contaminated and therefore is unusable – efforts can be made to stop contamination of freshwater sources, accessible water can be filtered, and rainwater can be harvested. Taking advantage of these opportunities requires multi-sectoral partnerships, bringing all stakeholders to the table, and collectively developing solutions to improve conditions that benefit all.

A family's health begins in the home, where access to potable water, adequate sanitation infrastructure and health habits ensure parents are able to work and bring home an income, and children thrive at school. Increased health results in decreased medical bills and increased economic capacity. This benefits the immediate community and region overall. Therefore, safeguarding homes so that they are places that support health, rather than compound or worsen health conditions, is critical. Adequate infrastructure, health literacy, and good hygiene and sanitation practices are all key for families to be healthy at home and have the opportunity to be active participants in their community's development. Fostering collaboration across disciplines and including communities and municipal leaders can bring about sustainable change and ensure that no one is left behind.

Appendices

Appendices

Appendix I

Letter from Rapheal Muma, Idenau municipality

A letter from Idenau municipality's local representative, Rapheal Muma, describing the current situation in Idenau and the value of the Needs Assessment, which forms the basis of this work.

Water and sanitation are global priorities. The United Nations Millennium Development Goals (MDGs) included the target to reduce by half the number of people without sustainable access to safe drinking water and basic sanitation. The Government of Cameroon included water and sanitation in its developmental agenda, "The Growth and Employment Strategy Paper". As a policy, the government of Cameroon plans by 2020 to improve the access rate to safe drinking water and basic sanitation infrastructure by 75%, by (i) rehabilitating existing infrastructure most of which was built more than twenty years ago; (ii) extending existing networks which have lagged behind urban expansion and population growth; and (iii) encouraging the realization of large-scale connection programmes.

The implementation of government hygiene and sanitation policy within local communities is better realized and managed by decentralized local authorities and civil society organizations operating at a grassroots level. Guided by the policy of decentralization and motivated to better the local communities especially for women, girls, and youth, we are being driven by this passion to make our environment better, healthier and liveable. We are committed to the fact that:

- We don't just live besides people, but we live with them.
- We don't just educate people but we learn from them.
- People don't just know us, they understand us.
- We are not all-knowing experts, but we are committed participants.
- We are not a replacement, but we are a facilitator.
- We do bring what we want but we ask what works for our communities.
- We don't have all the resources needed to provide for our community, but we strive to link our community with potential partners.

Our primary objective is to introduce cost effective domestic and industrial waste management systems and the development of viable alternative sources of livelihood for women and youths. The specific guiding objectives are:

1. Reduce household pollution and vulnerability of community members to infections to at least 40% from poor waste management in Idenau;
2. Increase access to potable water from 40% to more than 90% in all the communities of Idenau;
3. Increase community awareness regarding the dangers of poor waste control and management through the introduction of education on sustainability to at least 75% of the communities within the Idenau municipality; and
4. Reduce the dependence on unsustainable waste management practices for income generation by women through the introduction of environmentally friendly alternative livelihoods with an expected rise of household incomes for at least 65% of poor families.

Our expected short- and long-term results are:

1. Within 24 months, at least 200 households will be free from the dangers of poor waste control and management.
2. 5 community-based waste management and control groups are established and empowered for effective waste control and management with monitoring and evaluation services provided by the Idenau council.
3. At least 5 neighbourhoods with over 5,000 inhabitants that have access to public toilet systems to solve the total lack of toilets for households and along the streets and beautiful beaches of the municipality.
4. More than 10,000 community members to be engaged in the Education for Sustainable Development (ESD) Program with a mastery of environmental protection and social and economic empowerment practices integrated into their daily lives.

Envisaged activities to achieve these results are:

1 *Provision of community potable water:* Safe water remains a dream for the people of the Idenau municipality. Within the community, many women and children go long distances in search of surface water. This water is sadly of poor quality. Families with pipe borne water also face challenges. The stand taps are sometimes dry for more than two months, and when they are flowing, the water quality is also poor. To safeguard the community from water borne diseases, we envisage drilling boreholes for communities without pipe borne water. We also plan to improve the water quality for communities who already have piped water networks. This will involve providing at least 70% of the population with sustainable systems of potable water through: a) drilling of boreholes of 60–70 mL; b) piping the borehole; c) building a filtration system at the bottom of the borehole; d) installation of hand pump system with soakaway pit; e) training of management committees; and f) rehabilitation of existing water networks.

2 *Installation of hygiene and sanitation systems/waste management through community initiatives*: More than 60% of uncontrolled and poorly managed waste in Idenau is from households while about 30% is from improperly disposed industrial discharges mainly from the Cameroon Development Cooperation (CDC) and from fishing, the major economic activity in Idenau. As an expeditious way of controlling both domestic and industrial waste, we plan to establish a waste collection and disposal system by installing waste disposal cans at major junctions. These will be used to collect household waste and these cans will later be managed by the local waste control and management groups that will be created. We intend to acquire tricycles on behalf of the local waste control management groups. These tricycles will be used in collecting waste from households daily. The refuse will be separated into organic and inorganic waste and placed into the large cans. The organic waste will be sold by the local groups at very low costs to gardeners in the area to fertilize their vegetable farms and the inorganic waste will be sold to recycling firms in Douala.

3 *Construction of modern pit latrines with 5 stalls each:* In the Idenau municipality, more than 75% of households lack toilets and as such, the majority of household members defecate around houses, on beaches, and in the streets, creating a health and environmental threat for the population. We envisage the construction of modern latrines/toilets within the municipality. These toilets are expected to reduce the spread of bodily waste which without such infrastructure makes many places in Idenau appalling. Individuals and households will have access to these community/public toilets free of charge while the local waste control and management group will ensure the cleaning and renovation of the toilets together with the Idenau Council Administration.

4 *Introducing Education for Sustainable Development:* The creation of long lasting and sustainable solutions for problems cannot be limited to the direct provision of services, but rather a better starting point is the transformation of the mindsets of people to change their habits. Following this premise, an important component of this project will be the introduction of the Education for Sustainable Development (ESD)

Model to Idenau which targets the education of community members on issues such as environmental protection, climate change, leadership, and entrepreneurship. The ESD component will prioritize the education of women, youth and indigenous people on sustainable waste management and control education, and its benefits to community resilience. The ESD component will run community and school based focused training for the generation of ideas and transfer of knowledge and for the management of local solutions to problems such as waste control and management within the Idenau municipality.

Additionally, Idenau municipality (A dream land for investment) is a natural tourist haven that is unfortunately undergoing environmental degradation. This scenic and beautiful natural landscape and its people need environmental restoration for both environmental and economic health.

~ Rapheal Muma

Appendix II

Needs assessment survey

Participant ID:__________
Date:____________ Participant Signature:____________________

To conduct this needs assessment, please read all questions that are not in italics out loud to participants. All italics are instructions for the interviewer. Signing above and beginning this survey implies that the participant consents to taking the survey. One person per household at least 18 years of age or older can complete this survey. Each participant has a unique ID number.

Interviewer reads to Participant: This 15-minute survey will help determine challenges & barriers regarding sanitation infrastructure in your community, and can act as a step forward in taking action to improve the community needs. Your signature on this survey indicates that you are consenting to participate in this needs assessment report. Please answer the following questions honestly & to your best ability.

1 The last time you needed to defecate, were you:

- At home
- Outside home (*specify*__________)

2 Did you use a latrine or toilet?

- No, I openly defecate [go to Q3]
- Yes, I use a latrine/toilet [go to Q2.1]
- Other (*please explain*______) [go to Q3]

2.1 Can you show me where your toilet or latrine is? Indicate what type of latrine it is:

- Flushing toilet
- Pit latrine
- Composting toilet
- No facilities (*circle one:* bush, field, beach, open land)
- Other (*specify:* _______________)

2.2 Are latrines or septic tanks emptied when they fill up?

- No
- Yes [*circle one*: manual or pumping pipes?]

2.3 Do you share this toilet facility with other households?

- No • Yes *(how many households:_____)* • Don't know

3 Are there children under 5 living in the household?

- No *[go to Q4]* • Yes *[go to Q3.1]* (*# of children:_______*)

3.1 The last time the child(ren) passed stools, what did you do to dispose of the stools? *Select all that apply.*

- Child used toilet/latrine
- Thrown into garbage
- Dumped into surface water (*circle:* ocean, river, dam, lake, pond, stream, canal, irrigation channels)
- Put/rinsed into toilet or latrine
- Buried
- Put/rinsed into drain or ditch
- Left in the open
- Don't know
- Other (*specify:__*)

3.2 If or when your child gets diarrhea, how does it get disposed? *Select all that apply.*

- Child used toilet/latrine
- Thrown into garbage
- Dumped into surface water (*circle one:* ocean, river, dam, lake, pond, stream, canal, irrigation channels)
- Put/rinsed into toilet or latrine
- Buried
- Put/rinsed into drain or ditch
- Left in the open
- Don't know
- Other (*specify:___*)

4 **How is solid waste (garbage) disposed of?** *Select all that apply.*

- Collected by municipal waste system
- Openly dumped
- Burned
- Don't know
- Buried and covered
- Other (*specify:*_______)

5 **Do you separate your garbage?**

- No *[go to Q6]*
- Yes *[go to Q5.1]*

5.1 Can you show me how you separate them? *Select all that are separated:*

- Plastic bottles
- Glass bottles
- Plastic bags
- Paper bags
- Paper or plastic containers
- Cans

6 **Please indicate whether you strongly disagree (SD), disagree (D), neutral (N), agree (A) or strongly agree (SA) with the following sentences** (*There is no right or wrong answer*):

6.1 Open defecation provides no privacy:________

6.2 Open defecation is bad:_______

6.3 People who defecate in the open put their children at risk of disease:____

6.4 People who defecate in the open put the whole community at risk of disease:____

6.5 Open defecation is common:______

6.6 In this community, a mobile phone is more important than improving a latrine:

6.7 In this community, school fees are more important than improving a latrine:________

6.8 In this community, a bicycle is more important than improving a latrine:______________

7 **Do you think latrines are physically safe to use?**

- Never
- Rarely
- Some of the time
- Most of the time
- Always

8 **Is potable water available to you?**

- No *[go to Q9]*
- Yes *[go to Q8.1]*

8.1 Is potable water available year-round?

- No *[go to Q9]*
- Yes *[go to Q8.2]*

8.2 How long is it unavailable?

- One day
- One week
- One month
- More than one month

8.3 What are the reasons for unavailability of potable water? *Select all that apply.*

- Damaged water sources
- Theft of parts
- Dried water sources
- Lack of skilled workforce to fix the source
- Non-availability of maintenance parts
- I don't know
- Other (*specify:*________)

9 Can you show me where you get your drinking water for members of your household? *Select all that apply. Indicate where they get their drinking water:*

- Piped water to yard/plot
- Dug well
- Bottled water
- Large-scale tank delivery
- Public tap/standpipe
- Surface water (*circle one:* ocean, river, dam, lake, pond, stream, canal, irrigation channels)
- Spring
- Small-scale tank delivery
- Off property/far away
- Tubewell/borehole
- Rainwater collection
- Other (*specify:*______)

10 Can you show me where you get the water used by your household for other purposes, such as cooking and handwashing? *Select all that apply.*

- Piped water to yard/plot
- Rainwater collection
- Surface water (*circle one:* ocean, river, dam, lake, pond, stream, canal, irrigation channels)
- Large-scale tank delivery
- Public tap/standpipe
- Bottled water
- Tubewell/borehole
- Small-scale tank delivery
- Off property/far away
- Dug well
- Spring
- Other (*specify:*__)

11 If you travel to your water source, how frequently do you visit your main water point?

- Don't travel to water source *[go to Q12]*
- times per week *[go to Q11.1]*

11.1 What means of transport do you normally use to get to the water point? *Select all that apply.*

- Walking
- Bicycle
- Cart
- Other (*specify:*________)

11.2 How long does the roundtrip from home to the water source take?

- Less than 30 minutes
- Between 30 minutes to 1 hour
- More than 1 hour
- Don't know

12 Do you have access to a handwashing facility?

- No *[go to Q13]*
- Yes *[go to Q12.1]*
- Don't Know *[go to Q13]*

12.1 Are both soap and water currently available at the hand-washing facilities?

- Yes, water and soap
- Soap only
- Water only
- Neither water or soap

13 When do you wash your hands? *Select all that apply.*

- Before and after eating
- After using toilet
- None of the above/don't wash hands
- Other (*specify:*__________)

14 When do you wash your children's hands? *Select all that apply.*

- Before and after eating
- After using toilet
- None of the above/don't wash child's hands
- Other (*specify:*______)

15 Do you treat your water to make it safer to drink?

- No *[go to Q16]*
- Yes *[go to Q15.1]*
- Don't Know *[go to Q16]*

15.1 What do you usually do to the water to make it safer to drink? *Select all that apply.*

- Boil
- Use a water filter (ceramic, sand, composite, etc.)
- Add bleach/chlorine
- Strain it through a cloth
- Don't know
- Solar disinfection
- Let it stand and settle
- Other (*specify:*_______)

For reference, "diarrhea" is liquid faeces that occurs when defecating.

16 Which of the following diseases have you suffered from in the past 2 weeks? *Select all that apply.*

- Diarrhoea
- Respiratory tract infections
- Dysentery
- Cholera
- Skin diseases
- Bilharzia
- Malaria
- Typhoid
- Anemia
- None of these
- Other (*specify:*____)

If there is no child(ren) in household under 5 years old, skip Q17 and go to Q18.

17 Which of the following diseases has your child(ren) under 5 suffered from in the past 2 weeks? *Select all that apply.*

- Diarrhoea
- Respiratory tract infections
- Dysentery
- Cholera
- Skin diseases
- Bilharzia
- Jaundice
- Malaria
- Typhoid
- Anemia
- None of these
- Other (*specify:*____)

18 What is your gender?

- Man
- Woman
- Don't know
- Other (*specify:*________)

19 How old were you on your last birthday?

- 18–24 years old
- 55–64 years old
- 25–34 years old
- 65–74 years old
- 35–44 years old
- 75 years or older
- 45–54 years old

20 What is your ethnicity?

- Arabes-Choa/Peulh/Haoussa/Kanuri
- Bantoi'de Sud-Ouest
- Beti/Bassa/Mbam
- Biu-Mandara
- Bamilike/Bamoun
- Kako/Meka/Pygmé
- Adamaoua-Oubangui
- Côtier/Ngoe/Oroko
- Foreign/Other(*specify.*____)
- Grassfields

21 What is your religion?

- Catholic
- Animist
- Protestant
- None
- Muslim
- Other (*specify.*_______)

22 What is your highest level of education?

- None
- Primary
- Secondary 1st cycle
- Secondary 2nd cycle
- Superior

23 What is your occupation?

- Agriculture (Farming/Fishing)
- Engineering
- Petty Trading (Business)
- Employee of house
- Teaching
- Nursing
- Other (*specify.*______)

24 What is your average monthly household income (in XAF)? XAF__________/month

25 Are you the head of the household?

- No *[go to Q25.1]*
- Yes *[go to Q26]*

25.1 What is your relationship with the head of household?

- Spouse
- Aunt
- Brother/sister
- Grandparent
- Uncle
- Other (*specify.*______)

26 How many people live in your home, including yourself?

- 1–2 people
- 7–8 people
- 3–4 people
- 9–10 people
- 5–6 people
- More than 10 people

27 Who do you think is the most influential group or organization to communicate about water, sanitation, and hygiene in your community? (*Choose one*)

- Village council
- School teachers
- Youth group
- Disaster preparedness committee
- Community nurses
- Member of women's union

End of Survey

Appendix III

Community discussion questions

The following questions are meant for group discussion. The community discussion will be conducted among the members of the community to understand the barriers and challenges they may face related to sanitation infrastructure. There are a total of 10 questions that should take approximately 30–45 minutes. All responses will be kept confidential and anonymous.

1 What are three of the biggest problems within your community?
2 Are there particular traditions concerning water, sanitation, or hygiene behaviour?
3 What do you think is the cause of diarrhoea? Can it be treated, and if so, how?
4 What do you think is the cause of malaria? Can it be treated, and if so, how?
5 Where is food stored in your home? What are the types of food eaten, and are they washed, cleaned, etc?
6 What is the one thing you would like to see improved within your community?
7 When was the last time you went to a health centre? What was the reason for going?
8 If you store water in your household, how do you store it? If in a container, is it covered? Where do you store it?
9 What are the locations where most people in your community defecate? Where do you usually defecate? Why do you choose this particular location?

What types of resources or training do you think would be beneficial for the community in terms of water, sanitation, or hygiene?

Appendix IV

Consent form for community discussion

Purpose: A Needs Assessment is being conducted in Idenau, Cameroon to better understand the sanitation infrastructure. This assessment will be used to identify, assess, and analyse challenges and opportunities in the area of disease transmission and prevention, local conditions, and traditional approaches to design and construction. The purpose of this infrastructure needs assessment is to detail the current state of the Idenau sanitation infrastructure, provide an overview of challenges currently being faced, and suggest solutions and infrastructure strategies.

Participants' Rights: The participant has the right to skip any questions that they do not wish to answer at any time during the discussion. All responses will be kept anonymous and confidential. This consent form will be kept separate from the data records to ensure confidentiality. The participant has the right to withdraw at any time during the discussion without penalty. Participating in this discussion is completely voluntary. If you have any questions, concerns, or complaints about your right as a participant, please contact Rapheal Muma.

Potential Risks and Discomforts: Questions asked during this needs assessment may be personal or private information. We ask that you please be honest with all answers provided. There is no right or wrong answer. Please be sure that your responses are as accurate as possible. To respect the privacy of others participating in the discussion, we ask that you keep the survey questions and responses to yourself and not share them with any other individual after the community discussion has been completed.

Potential Benefits: This needs assessment is designed with the community in mind. The responses you provide are intended to help the development of a community-wide improvement project. We hope that, in the future, a sanitation infrastructure will be implemented in Idenau to help control waste and benefit the health of those living in Idenau. This community discussion will be a first step towards the potential improvement in Idenau's sanitation infrastructure.

Confidentiality: We will do all that we can to make sure there is no loss of confidentiality. Your name will not be included in any forms other than this informed consent form. This informed consent will be stored securely in *************. All participants will be de-identified and given a participant ID to ensure anonymity of the responses.

All electronic information (including any digital audio recording) will be kept in passcode protected computer files and will only be seen by project staff. All data collected by paper will be stored in filing cabinets and will only be seen by research staff.

If we write a report or article about this project, all results will be reported in aggregate, meaning that we will not report any information about any one person. If you are interested in seeing these results, you may contact **********. Your information may be shared with representatives of Idenau, Cameroon or governmental authorities if you or someone else is in danger or if we are required to do so by law. Please sign below if you agree to be audio recorded during the discussion.

Statement of Consent: Your signature indicates that you are at least 18 years of age, you have read this consent form or have had it read to you, your questions have been answered to your satisfaction and you voluntarily agree to participate in this needs assessment. Your signature acknowledges receipt of a copy of this consent form as well as your agreement to participate.

If you agree to participate, please sign your name on the sheet provided.

Name:______________________ Date:______________________

Signature:______________________

Index